4,00

D1459826

J. M. W.

THE DISCOVERY OF OUR GALAXY

ALFRED A. KNOPF

THE DISCOVERY
OF OUR GALAXY

CHARLES A. WHITNEY

NEW YORK   1971

Dedicated to my father,
Charles Smith Whitney,
whose life was a work of art

CONTENTS

ILLUSTRATIONS

PROLOGUE

For two thousand years, some have known the earth to be round. But only for twenty years have we known the detailed shape of our galaxy—the spatial implications of that diffuse band of light we call the Milky Way. This book retraces the trail of ingenious arguments by which astronomers have interpreted the sky and built a picture of the galaxy in which we spin.

On a clear, moonless night in midwinter or midsummer, a plume of starlight rises motionless behind the scattering of constellations. To our eyes, the Milky Way is colorless, but we are deceived by the faintness of the light. If our eyes were more sensitive we would see a splendid display of colors: the blue and red of Orion's stars, the deep yellow of Arcturus, the silver of Spica, and the sunlight yellow of Castor and

Pollux, glowing wisps of red and pale blue, an occasional gap of dull red in the banks of faint stars.

The Milky Way is our island universe—our galaxy—and its shape had been outlined by inspired guesswork before 1800, but only within the past two decades have the arguments been made convincing and quantitative. New tools were required at each step, and many tools relied on new concepts, not merely new techniques.

Yet, even lacking quantitative tools, astronomers might have made substantial progress much earlier if it had not been for one disastrous fact, overlooked until 1930. Interstellar space is filled with obscuring dust that weakens starlight, falsifies estimates of distance, and totally blocks the visible light of vast portions of the Milky Way.

The recognition that our Milky Way is a spiral galaxy depended heavily on irrational elements of man's nature: his quest for divine revelation in the sky; his ability to unbridle his imagination and face ridicule from those he most respected; his stubborn insistence that for some problems a wrong answer is better than none; his willingness to hold two mutually contradictory ideas in his head and work with them both; his faith that theories should be esthetic; his preoccupation with the problem of his own identity.

It is just this appearance of irrationality that makes the work of astronomers so fascinating today and that provided the peaks of scientific excitement in the past. The discoveries of the great astronomers were no more inevitable in their time than the paintings of Rembrandt were in his. What they wrote, what they sensed in themselves, stood at the tip of their perceptions.

Taxonomy of the sky—classification and identification of the varied objects detected by telescopes—was the first major step toward identification of the Milky Way. Bright wisps of nebulosity played a confusing role; they appeared in a variety of guises, as though an actor had changed his costume between every speech. Some astronomers thought these patches were composed of faint stars, others thought they were glimpses of a shining fluid filling the outer reaches of space— illuminations of the ether. Some thought they were fragments of our own Milky Way, others that they were distant Milky Ways. The evidence was discussed endlessly; most of it was circumstantial and unconvincing, even to the proponents. The spectroscope permitted analysis of the light of nebulae toward the end of the nineteenth century, and the answer was "yes and no" to all the earlier questions. For a while the arguments became even hotter.

Slightly less than fifty years ago, the elements of the pattern of the Milky Way began to emerge: Shapley located the center and set the earth and sun near its periphery. The rotation of our Milky Way was discovered, and Hubble showed that some of the tiny spirals of light must be vast star systems like our own.

Man at last had a perspective on his own situation. Our galaxy had been discovered.

Many questions remain unanswered—and still more remain unasked—but we have begun to understand the aging of these swirls of stars, and there is some evidence that the chronology of the universe may be traced back to a flash of light. The ancient Hebrew tradition "God said, 'Let there be light,' and there was light" appears to be the best modern description of the genesis.

Cosmic violence has recently been discovered in quasars and exploding galaxies; even the center of our own galaxy is a seat of violence. Whether these are birth pangs or the throes of death must remain unknown while we await more measurements and some good ideas.

C. A. WHITNEY

Weston, Massachusetts
January, 1971

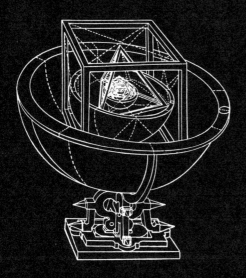

1 ANCIENT SPECULATION: PRELUDE TO SCIENCE

When I try to imagine the sky of the Ancients, I see a full moon rising at sunset, turning from orange to white and bathing the forest in silver light. A hunter rests for a moment at midnight, and he notices that the moon has again turned orange. As he watches, a dark red shadow slides across its face; he can no longer see well enough to hunt; the constellations stand out more clearly in the darkened sky, but he knows the moon will return and he is not afraid.

Once a year or so, when the moon had vanished as a slim crescent in the morning sky, a peculiar affliction overtook the sun: it imitated the moon. This would first be noticed among the spots of light beneath the trees—no longer round, they became thin crescents and as they grew thinner the sun's light weakened as though passing through a thick cloud of smoke. Looking up, the Ancients saw that the sun itself had taken the shape of the moon and was about to vanish as the moon had vanished a few days earlier. But then, the crescent rotated and began to enlarge until the full round disk had been restored. The affliction had passed.

One day, perhaps twenty years after the hunter had first seen a partial eclipse, an unheralded terror descended from the daytime sky. In the midst of a commonplace eclipse, when the sun's crescent had become unusually thin, a curtain of night rushed across the sky. The sun was plucked from its socket, leaving a black hole rimmed with silver and blue; the horizon flared in orange. Stars unseen for months reappeared, and birds sang their evening songs; women sheltered their children.

Then the sun returned. A drop of light expanded about the socket's edge and the silver crown faded into the dazzle. After sunset, familiar stars reappeared to mark the slow roll of the night and the subtle progression of the seasons.

Thales of Miletus, in the sixth century B.C., marked down all these traumatic days and invented a geometrical theory that would reproduce their pattern. He thus rid the world of its greatest terror: the total solar eclipse.

Among the ancient Greeks, speculation on nature was partly self-protective, but it was also seen as an act of heightened perception: *to theorize* was *to behold*. The need to test theories was no more evident then than the need to test a poem is now. For example, in the first century B.C., the Roman writer Lucretius said of his theories, "One of these must be the right one, and its force applies to our own earth as well as starry spheres; but no one, plodding forward step by step, as I do, dares to say which one is true." Lucretius claimed that the fear of death was the greatest cause of unhappiness, and that religion only aggravated the fears. He wrote to assure men of the mortality of their souls and of the blissful ignorance they would enjoy after death. He was quite explicit in describing his motives:

> *Come, come,*
> *Listen to what is left, and hear more clearly:*
> *I know how dark and difficult things are,*
> *But the high hope of praise has moved my heart*
> *As if by the wave of a wand, and deep within*
> *Filled me with sweet devotion to the Muses. . . .*

> *. . . I teach great things,*
> *I try to loose men's spirit from the ties,*
> *Tight-knotted, which religion binds around them.*

Thus, through rational thought, Lucretius hoped to follow Epicurus, that "glory of the Greeks," for once his

> *. . . divining sense,*
> *Begins its proclamation, telling us*
> *The way things are, all terrors of the mind*
> *Vanish, are gone; the barriers of the world*
> *Dissolve before me, and I see things happen*
> *All through the void of empty space.*

But in their eagerness to assure themselves that the only reality was the reality of the senses, that the only judgment was the test of pleasure and pain, the Epicureans lost their grip on the universe. They insisted that the human senses were infallible, and that things are just the way they appear to be. They admitted errors in thought, but lapses of perception were ruled out.

Lucretius obliterated the distinction between appearance and reality in the sky, saying:

> *The sun must be as large, as hot, as it*
> *Appears to us, no more, no less. . . .*
> *So, since our sense feels the flooding heat*
> *Of sunlight's comforting appearances,*
> *Our apprehensions of that form and size*
> *Must be correct; there's nothing you can add,*
> *Nothing subtract. The same way with the moon—*
> *Whether its light is borrowed or derived*
> *From its own substance, makes no difference—*
> *The moon can be no larger than it seems*
> *To our watching eyes.*

The following lines may provide the clue to the origin of the idea that the sun and moon are just as we see them:

> *Whatever we see, far off*
> *Is blurred before it shrinks. And so the moon*

Affording us an outline sharp and clear
Must have the same dimensions in the sky
As we down here observe.

Lucretius held that the outline of an object becomes blurred with distance before the apparent size of the object is diminished. In the days of perspective drawing and photography this is nonsense, but I think we may interpret it more charitably. When we look at a row of trees shrinking in the distance, we know that the distant trees seem smaller—that is, they cover a smaller portion of the sky. At the same time we are aware that the tree does not shrink in physical size merely because we walk away from it.

The distant trees also differ in the amount of detail they reveal of their twigs and leaves. So, if we are careful observers, we correlate the apparent smallness of a tree with the loss of discernible detail in its structure.

I think this is the point at which the Epicurean argument, as relayed to us by Lucretius, fails: he assumed that his eye could see all there is to be seen on the face of the moon. The possibility of a host of unseen details on the moon did not enter his mind, and was in fact inconsistent with the contemporary view of the moon as a pristine celestial sphere.

The astronomical consequences of Epicurean thought were quite dull: the sun and the moon are just what they look like, and even the stars "can hardly be much less or greater than they seem to us." Stars were merely "stars." A few simple experiments might have shown the Greeks that their philosophy was a very fragile vessel, but it is not man's aim to destroy his own philosophy when it has carried him into calm waters.

But what are we to make of such men as Democritus, of the sixth century B.C., who imagined the universe to be populated with an infinity of worlds like our own? Twenty-five centuries passed before the slightest evidence directly supported that possibility. Shall we call him a dreamer or a brilliant philosopher? Whatever we claim for Democritus and others like him, we must admit that he was out of touch with his time. If he deserved to be heard because he was interesting, he also deserved to be called irrational.

What is the role of such a man in history?

If we say that he demonstrates that there is "nothing new under the sun"—a quotation older than Democritus—we are telling only part of the story. While it may be true that many of our theories are resurrections of ancient imaginings, it is also true that we have adopted new attitudes and new approaches toward the old ideas; we have found new modes of correspondence between the external world and the products of our mind. In short, we have invented science. In this sense, if they accomplished nothing else, men like Democritus helped mark the flow of history like harbor beacons seen far astern.

But I think they have played another role which is more crucial: they have given us the courage to permit the fading colors of our own rationality to be tinted by the brilliance of intuition. Democritus did for astronomy what religious prophets like Jesus and Mohammed did for ethics and morality. These men may have sounded like lunatics on a mountain, but they spoke from the deepest boundaries of their rational selves. They pushed the boundaries of sanity slightly wider and made room for creativity.

2 FROM ANCIENT TO MODERN SCIENCE THROUGH SCHOLASTICISM, PARANOIA AND MARTYRDOM

✡

On my desk is a book in which the history of astronomy from the third to the fifteenth centuries is allotted two blank pages.

The first glimmerings of scientific rebirth after the Dark Ages appeared in the fifteenth century within the Roman Catholic Church. Nicholas Cusanus, a cardinal, was not an astronomer, but his expanded vision turned men once again to an intellectual exploration of the sky. Cusanus must have been an optimist: he proposed a remodeling of the Church and of the lives of individual men. His book *Of Learned Ignorance* is an elaborate scholastic discussion of the nature of God. But in the book, we sense a modern approach to science. Cusanus had no fear of ignorance; he was not confined to the limited universe of the Greeks; he wrote with the air of an enthusiastic explorer, confident and exhilarated to be on the high seas. Ignorance was a promise of future delights, an exotic island, rather than a cavity to be filled with speculations.

Cusanus proclaimed the infinite variety of the universe:

> In the universe there is nothing that does not enjoy a certain singularity that it shares with no other. . . . Individuating principles do not meet in precisely the same harmonious proportion in one thing as in another.

In the mind of Cusanus, this individuating principle had two consequences. First, the measurable world presents an infinitely rich challenge to rational thought—there is no limit to the "knowable." Second, each man is unique, and the cardinal believed that man should therefore be "content with the manners of his own people, with his own tongue and all else of his home, and he should find something peculiarly dear to him in the soil that gave him birth. In such spiritual ground there grow unity and peace without envy, as far as possible here below."

These simple tenets have become the *sine qua non* of modern science. A man who believes that he can empty out the jar of the universe and discover all of its secrets is insufficiently humble to engage in modern science. On the other hand, a man who remains unconvinced of the uniqueness and value of his own experience will lack the ego of a scientist.

The first astronomical scientist of any consequence after the Dark Ages was Nicholas Copernicus, born in 1473 when man's view of the universe was still fashioned on the ideas of the ancient Greeks and Egyptians. The sky was a vast machine, a concatenation of shells carrying the sun, the planets, and the stars about the head of man. The universe swam in the breast of God.

Copernicus possessed the humility and sense of uniqueness that Cusanus had offered modern man, but he lacked a third principle we now attribute to scientists—the urge and the ability to communicate. It could be argued that Copernicus's utter failure to communicate led to his rather peculiar role in the history of science.

Copernicus was not wholly in either the ancient *or* the modern world, nor was he truly outside both. He set the sun at the center of the

net, in quo terram cum orbe lunari tanquam epicyclo contineri
diximus. Quinto loco Venus nono menfe reducitur. Sextum
deniq̃ locum Mercurius tenet, octuaginta dierum fpacio circũ
currens. In medio uero omnium refidet Sol. Quis enim in hoc

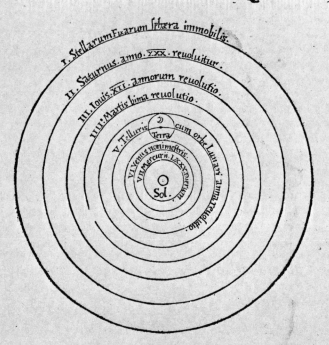

pulcherrimo templo lampadem hanc in alio uel meliori loco p●
neret, quàm unde totum fimul pofsit illuminare? Siquidem non
inepte quidam lucernam mundi, alij mentem, alij rectorem uo●
cant. Trimegiftus uifibilem Deum, Sophoclis Electra intuentẽ
omnia. Ita profecto tanquam in folio regali Sol refidens circum
agentem gubernat Aftrorum familiam. Tellus quoque minime
fraudatur lunari minifterio, fed ut Ariftoteles de animalibus ait,
maximam Luna cum terra cognatiõe habet. Cõcipit interea à
Sole terra, & impregnatur annno partu. Inuenimus igitur fub
hac

The solar system, according to Copernicus. Reproduction of a page from
the 1566 edition of Copernicus's *De revolutionibus coelestium*. The sun
is at the center, and the planetary orbits are indicated, as well as that of
the moon about the earth. The Copernican model, when first developed,
was merely a computational device rather than a coherent physical theory,
so the actual distances of the planets were unimportant. In this figure,
they are far from correct. (*Yerkes Observatory, University of Chicago*)

solar system. He stopped the stars and set the earth into daily rotation, thus opening the possibility that the stars might be distant suns. Yet his book, *The Revolutions of the Heavenly Spheres,* published in 1543, was "an all-time worst-seller" in the words of twentieth-century writer Arthur Koestler; even the mere thousand copies of the first edition were never totally sold. The book was dullness epitomized, and it was neglected for three-quarters of a century; yet it has become the focus of what we know as the "Copernican Revolution," the transition to the modern view of the universe.

Copernicus appears as an arch-conservative in science. He attempted to restore ancient physics by constructing a model of the solar system which would fit the observations of planetary motions available to him (most of them quite ancient). His mind was wedded to the ancient ideal of circular motions, and, in the words of Kepler in the seventeenth century, "Copernicus tried to interpret Ptolemy rather than nature."

We still do not know for certain whether Copernicus thought his creation was more than a mathematical device. It is difficult to imagine that he did not take it seriously, but Copernicus was an enigmatic character, and Koestler concludes, "It was perhaps this doubt about the real value of his theory which broke his spirit."

Copernicus's book was not published until the year of his death— he is said to have died with the first copy in his hands—and this was almost four decades after he had first worked out the model.

Orphaned as a child, with a leprous brother, Copernicus apparently developed a paranoid sensitivity to the scorn of his contemporaries. I can well imagine that he wished he had never started, that he would like to have forgotten the entire venture. He had sought to tie together man's view of the world, to tuck in the tails and tighten the weakened seams of the ancient tradition; but as he worked he discovered to his horror that he had to replace the traditional model. His aim was being thwarted before his eyes—the system he had hoped to bolster was destroyed by his own hand! Where would he find the gratitude he had sought? Certainly not in this book: The book must not be published, yet he could not bring himself to destroy it.

Much of the foregoing is speculation, but the Preface to *The Revolutions* is a unique confession and perhaps it "explains" the enigma of Copernicus:

TO THE MOST HOLY LORD, POPE PAUL III

The Preface of Nicholas Copernicus to the Books of the Revolutions

I may well presume, most Holy Father, that certain people, as they hear that in this book about the Revolutions of the Spheres of the Universe I ascribe movement to the earthly globe, will cry out that, holding such views, I should at once be hissed off the stage. For I am not so pleased with my own work that I should fail duly to weight the judgment which others may pass thereon . . . yet I hold that opinions which are quite erroneous should be avoided.

Thinking therefore within myself that to ascribe movement to the Earth must indeed seem an absurd performance on my part to those who know that many centuries have consented to the establishment of the contrary judgment, namely that the Earth is placed immovably as the central point in the middle of the Universe, I hesitated long whether, on the one hand, I should give to the light these my Commentaries written to prove the Earth's motion, or whether, on the other hand, it were better to follow the example of the Pythagoreans and others who were wont to impart their philosophic mysteries only to intimates and friends, and then not in writing but by word of mouth. . . . In my judgment they did so not, as some would have it, through jealousy of sharing their doctrines, but as fearing lest these so noble and hardly won discoveries of the learned should be despised. . . . Reflecting thus, the thought of the scorn which I had to fear on account of the novelty and incongruity of my theory, well-nigh induced me to abandon my project.

These misgivings and actual protests have been overcome by my friends. . . . They insisted that, though my theory of the Earth's movement might at first seem strange, yet it would appear admirable and acceptable when the publication of my elucidatory comments should dispel the mists of paradox. Yielding then to their persuasion I at last permitted my friends to publish that work which they have so long demanded.

I doubt that the history of science would have been much different if Copernicus had published his book a century earlier. In his time the world of science lay dormant, and even if the book had been delightful to read it might well have been ignored.

Toward the end of the sixteenth century, Giordano Bruno aroused the groggy world, asking it to fling its mind far beyond the planets. He speculated that the cosmos extended to infinity—without an edge. After all, he asked, if space stopped, what came next? with what was it then filled? how thick was the wall at the edge of the universe?

This in itself was not so shocking; but Bruno went considerably further—he postulated a multiplicity of worlds: suns and planets with life, unseen companions for the race of man. He toyed with man's conception of himself; for this, and for magical claims and political entanglements, he was burned in 1600.

The historian Francis R. Johnson has recently pointed out that Thomas Digges, an English scientist, also deserves credit for spreading the idea of the infinite universe. He says Digges "was the first modern astronomer of note to portray an infinite, heliocentric universe, with stars scattered at varying distances throughout infinite space. . . . In Digges's book the average Englishman, who lacked the opportunity or the learning to read and understand Copernicus's great work in the original, got his first authoritative exposition of the new heliocentric theory and the arguments in its favor. . . . It was Digges's treatise proclaiming the idea of an infinite universe in conjunction with the Copernican system, and not Bruno's speculations on this subject, that influenced the thought of sixteenth-century England. Digges, moreover, was an eminent scientist, and was eager to verify his ideas by experimental methods. Bruno, on the other hand, had arrived at his notions entirely through metaphysical speculations."

Johnson's remarks are not totally convincing, because I do not see the relevance of Digges's "experimental methods" to the question at hand—not in the sixteenth century at any rate.

The following passages indicate the spirit of Bruno's writing. They are taken from his book *On the Infinite Universe and the Worlds:*

ELPINO: How is it possible that the universe be infinite?
PHILOTHEO: How is it possible that the universe be finite?
ELPINO: How do you claim that you can demonstrate this infinitude?
PHILOTHEO: Do you claim that you can demonstrate this finitude?
ELPINO: What is this spreading forth?
PHILOTHEO: What is this limit? . . . If the world is finite and if nothing lieth beyond, I ask you WHERE is the world? WHERE is the universe? Aristotle replieth, it is in itself. . . . [But] position in space is no other than the surfaces and limit of the containing body, so that he who hath no containing body hath no position in space. What then

dost thou mean, O Aristotle, by this phrase, that "space is within it-self"? What will be thy conclusion concerning that which is beyond the world? If thou sayest, there is nothing, then the heaven and the world will certainly not be anywhere.

FRACASTORO: The world will then be nowhere. Everything will be nowhere.

PHILOTHEO: The world is something which is past finding out. . . . If thou sayest that beyond the world is a divine intellect, so that God doth become the position in space of all things. . . . I say [that it is] impossible that I can with any true meaning assert that there existeth such a surface, boundary or limit, beyond which is neither body, nor empty space, even though God be there. For divinity hath not as aim to fill space, more therefore doth it by any means appertain to the nature of divinity that it should be the boundary of a body.

There were two specifically astronomical corollaries to Bruno's idea. First, the stars must all be suns and their faintness must be caused by great distance. This idea was not new, and Bruno brought no fresh evidence. But there is a second corollary which appeared to be new: If the universe has no edge, it cannot have a center; it cannot have an up and down, or right and left. How then can we make a map? How can we determine our location in the universe? Presumably we could measure the distances to the nearby stars and plot them on a three-dimensional array, and then by triangulation we could fix our position. If we move off the edge of the map, we can make a new one to extend the old. But if we must count our way from star to star our map-making will become tedious, as though today's highway maps were built to show every tree.

Suppose we try to simplify a highway map using natural landmarks. We would choose large objects, such as lakes and mountain ranges. And if we wish an even simpler map covering more territory we would use the borders of the continents and we might not show the lakes at all. In this way we could build a map of the world which could be read at a glance—because we have left the trees off.

Can we transfer this technique to the universe? Bruno said, "No, we cannot make a continental map of the universe." There was no large-scale structure in the universe—all was uniform.

The loneliness conveyed by Bruno's idea is undeniable—man was no longer uniquely centered in creation.

3 THE SKY AWAKES

Shortly before Bruno's death in 1600, Tycho Brahe made the first announcement of a "new" star in the sky. A few years later he observed a comet, and proved that it moved among the planets; thus he shattered the crystalline spheres which had been supposed to carry the planets and stars about the heavens. Other men had suggested that there were no crystalline spheres, but their arguments were speculative and uncompelling. It was left to Brahe to show that the question could be answered by looking up at the sky and recording what was seen night after night. He was the first to prove that bodies can move freely through astronomical space and that the sky can change.

A "new" star, or "nova," is not a new star; it is a new appearance of an old star; it is a brief flash—a violent explosion—that increases the brightness of the star by ten thousand- to a million-fold for a few months. Amazingly, the star gradually returns to a state not very different from its initial state—sometimes undergoing wild fluctuations as it fades. We have theories, but no definite clues to the cause of the explosion; no one has predicted a nova.

Novae are not quite so common as bright comets, but they are no longer considered an extremely rare occurrence. Yet, in all history before Tycho, only one nova appears in written records: the nova of A.D. 1054. We can only speculate that, until Tycho's day, no one had expected to see changes in the starry background, so no one saw any. Many comets, on the other hand, had been observed, but Aristotle had written that they were beneath the moon; he believed they flitted through the fringes of our atmosphere, and few people had doubted him.

Tycho Brahe, a noble Dane born in 1546, was a bit of an odd duck. He had acquired a common-law wife from among the servant class, and to explain this his biographers variously accuse him of wanting to spite his noble friends and of trying to simplify his life to leave room for astronomy by pairing with a simple soul. Another possibility, it is said, is that he despaired of finding a noblewoman willing to marry him, because he had lost most of his nose in a duel and sported a silver and gold replica stuck in place with glue. (When his well-preserved body was disinterred in 1901 by curious citizens of Prague, the nose was found to be missing.)

At the age of twenty-five he wrote an epitaph for his stillborn twin brother. Whether this inscription reveals an underlying guilt at his own existence, or whether Tycho was simply a prankster, I will not inquire. The fact is that the unnamed brother's tombstone was engraved with the Latin equivalent of the following:

> I who am dead have been resurrected. I was buried in this earth before I was born. Guess then who I was? I was unborn in my mother's womb, when death became my door to life. There was another enclosed with me, a brother who still lives, for I was a twin. God granted him a longer life than me, so that he might see the strange things on the earth and in the heavens. My fate has not been worse than his. He lives on earth, but I in heaven. He is subject on earth to a thousand perils, that happen on land, at sea, and in the stars. But I am with God in heaven where I enjoy everlasting peace and joy.

When he finally rests his tired limbs under the cold earth, then we will both be united in heaven and he will participate in everlasting joy. Until then he will have to bear patiently the body's burden and not envy me my joy. Owing to my sudden death I was denied a name among the living. He instead of me bears the name of my grandfather. My grandfather's name was Tynge and his surname Brahe, and my brother also is named Tynge. It is he who has honored my grave with these verses. He is now twenty-five years old.

When he was fourteen, Tycho witnessed a partial eclipse of the sun. This was not a particularly awe-inspiring sight in itself, but the accurate prediction of an eclipse would be an impressive accomplishment. To Tycho, it was "something divine that men could know the motions of the stars so accurately that they could long before foretell their places and relative positions." He procured the best almanacs available and began his own study of the planets—but was soon disappointed and perplexed. With a simple cross-staff, he measured the planetary positions and found the almanacs to be in error by several degrees. The discrepancy was too large for him, and he thought that he might improve man's knowledge of these heavenly motions. This became his driving motivation for the remainder of his life.

Tycho's fame throughout Scandinavia and Europe was established by a chain of events that began on the night of November 11, 1572. He had left his chemical laboratory, constructed in a shed provided by his uncle, and was winding his way across the darkened abbey where they lived, when he glanced up at the sky, probably as casually as any of us when we scan the sky in the evening. There, near the sprawling "W" of Cassiopeia high overhead, was a bright interloper making a parallelogram with the right-hand end of the constellation. He couldn't believe his eyes—a new star! Or was it his imagination? Stopping, he asked the servants who accompanied him if they saw it. Yes, they did. And he asked a group of peasants passing in a wagon the same question. Yes, it was undoubtedly there. Supper was delayed that night while he measured the distance of this star from the stars in the constellation of Cassiopeia.

Tycho had no telescope (he was the last great astronomer to die before Galileo turned his telescope on the sky), but he knew this object was no comet, because it had no tail; he knew it was no planet, because it twinkled strongly as only stars do; he knew it was far beyond the

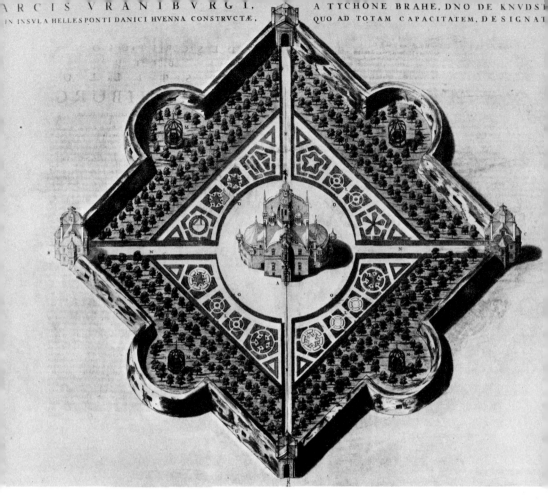

Uraniborg, Tycho Brahe's observatory, about 1584. This was the greatest observatory in the Western world. Telescopes had not yet been applied to astronomy, so measurements at this observatory were accomplished with devices resembling gun sights. (*Yerkes Observatory, University of Chicago*)

moon, because it remained motionless against the background of the other stars.

Astrologically, the nova presented an interesting problem because there were no rules for its interpretation, yet it must have portents. Tycho made up his own rules, and included predictions with the observations: As the star was initially like Venus and Jupiter, its effects would at first be pleasant; it then became like Mars, so troubled times were to follow; finally it was Saturnine, thus implying a time of death. Tycho's analyses were based primarily on the color of the star, and they suggested a middle stage of reddish color. Modern observations of supernovae confirm this change.

Even today, Tycho's star would be an astonishing sight; it was far brighter than the brightest planets—probably bright enough to cast visible shadows on a dark night—and it could be seen in daylight. No remnants of this explosion have yet been discovered, although there is a faint star in the proper spot. This may be the core of the original star.

(Another explosion, this one in A.D. 1054, has, however, left a spectacular vestige: the Crab Nebula. The filamentary structure of this nebula is quite prominent in the red light of hydrogen, and it is so fine and sharp that the outward motions of the gas can be detected on photographs taken only a few decades apart. Measurement of these visible motions across the background of the sky and comparison with the outward speed of the gas determined with a spectroscope have permitted an accurate determination of the distance of the nebula— 3600 light years, which puts it closer to the sun than most stars of our Milky Way.)

Five years after his discovery of the nova, Tycho was fishing in a pond on his private island, when he noticed another brilliant object in the evening sky. Venus, the brightest of the planets, was then a morning star, so Tycho's curiosity was aroused. When the sun had set and the sky had darkened, a long broad tail stretched across the sky—a comet.

For two months, Tycho carefully recorded the motions of the comet, and he became convinced that Aristotle had been wrong: the comet was not a "fiery meteor of the air"; it lay well beyond the moon. In fact, judging from its motion, the comet passed among the planets— it penetrated the crystalline spheres imagined to maintain the planets, and it thus shattered another legend. Literally hundreds of books about the comet were published within the next decade, and a new branch of astronomy had sprouted: the motion of comets.

The world had to wait over a hundred years for Newton to provide the key to the motion of comets, but the tracking of these evanescent bodies added zest to astronomy. Astronomers no longer had to confine themselves to the motions of the planets and the moon.

And what do we have now, four hundred years later, besides novae and comets? There are stars in pairs, and more complex groups in which orbital motions have been detected—but they are hardly lively.

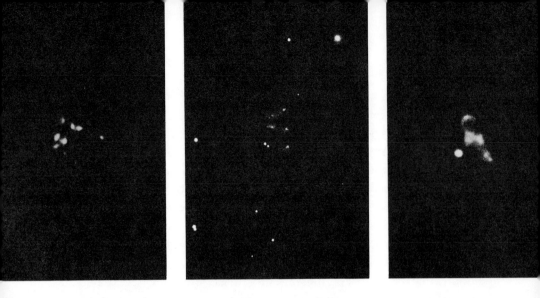

Young stars and nebulosity. These three small clusters are evidently in the process of rapid evolution from distended dust and gas clouds to stars, because pictures taken at intervals of several decades show unmistakable signs of condensation. (*Lick Observatory, University of California*)

There are stars whose light surges and sinks in periods from minutes to years. There are mobile clouds on Mars and Jupiter, but they conceal more than they reveal.

Most stars, the sun among them, stretch their lives over billions of years, so they appear the same from one century to the next, although some stars burn out within tens of millions of years.

But there is one recorded example of a star—or what certainly appears to be a star—being born within two decades. On an early photograph there is an amorphous nebula; on a later photograph a bright knot has developed in the nebula—a "new star" in the truest sense. George Herbig, the man who reported this condensation, maintains that it is not altogether incredible to imagine that we are watching the birth of a star; he and others had found many examples of faint stars deeply buried in nebulosity (plate, above) and these, too, may be the seat of star formation. Aside from the suggestive appearance of these nebulous stars, there is evidence of a flickering activity that implies instability and the accretion of the material to the outer layers of the stars.

There are also nebulae that vary, and the variations appear to be excited by a nearby star. One nebula looks like a comet, and the star at its "head" excites the gaseous tail to luminescence; as the star varies, so does the nebula (plate, below).

One spectacular example of life in the sky is the wave of light that flashed out from a nova discovered in 1901, in the constellation Perseus. The top plate on page 22 contains two photographs taken at an interval of seven weeks which show the circular nebula that started expanding from the star at the time of the outburst. The distance to this star can be

Hubble's variable nebula (N.G.C. 2261). During the nineteenth century, this nebula was discovered to vary erratically in brightness; in 1916, Hubble made it the object of one of his first researches. This is the first official photograph made with the 200-inch Hale telescope on Mount Palomar. (*Hale Observatories*)

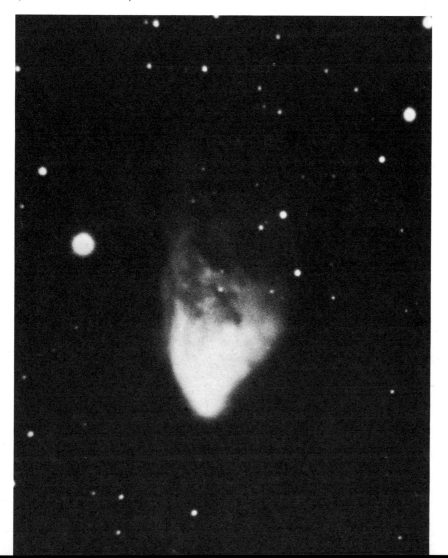

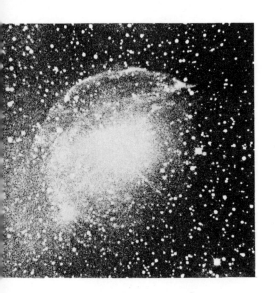

Reflection of light from the outburst of Nova Persei 1901. This star erupted on February 22, 1901. The left-hand photograph, taken on September 20, 1901, shows what appears to be a circular nebulosity. In the right-hand photograph, taken on November 13, 1901, the nebulosity has grown visibly. The expansion was so rapid that astronomers concluded they were seeing the expansion of a wave of illumination through a pre-existing nebulosity. The nebulosity ejected by the star only appeared years later; it is shown in the plate on page 24. (*Yerkes Observatory, University of Chicago*)

The changing brightness of the variable star RZ Cassiopeia. For purposes of measurement, the telescope was put slightly out of focus and the photographic plate was moved 1/20 of an inch between exposures. Each star produced a series of images, and the total time span was about three hours. This is an eclipsing variable star, in which one member of the pair alternately covers the other. (*Yerkes Observatory, University of Chicago*) inferred, and the outward speed of the nebula, judged from its expansion, is calculated to be just the speed of light—186,000 miles per second. There is only one explanation for this behavior: no nebula can be accelerated to such speeds by a nova explosion; we must be seeing the wave of light spreading out from the burst. It is as though we were to watch the nose of a searchlight beam dart through the air when it is turned on. (At the time of the burst, gas was ejected as well as light, and recent photographs indeed show a nebulosity expanding with a speed of several thousand miles per second—a hundred times slower than light.) (See plate, page 24.)

Every star and planet, when examined closely enough, has revealed variations in a number of ways. The atmosphere of the sun, for example, seethes with a filamentary activity, most of which is quite diffi-

The outburst of a nova. These two photographs depict the same portion of the sky; the arrow points to the image of the star before it exploded. A nova is not literally a new star, but its brightness temporarily increases by a factor of many thousands. This is Nova Herculis 1934, and its brightness declined to its pre-explosion value within a year. Recently the star has been discovered to consist of a very close pair of stars, and it has been speculated that an explosion in one of the stars may have been triggered by the presence of the other star. (*Yerkes Observatory, University of Chicago*)

Expanding nebulosity around Nova Persei 1901. Photographed with the 200-inch telescope in 1949, the nebulosity appears to be fainter than the star, and it displays a complicated filamentary structure. The apparent sizes of the stars in this photograph are caused by their differences in brightness rather than size. All stars are too distant to display their actual sizes in such a photograph. (*Hale Observatories*)

cult to detect, even from the earth. Tiny bursts last hardly more than a few seconds, but they produce emissions of enormous intensity; the solar surface heaves gently, very much like swells on the open ocean; sunspots emerge and grow over periods of months; the number of spots seen on the sun increases and decreases with a period of eleven years, and the magnetic field of the sun reverses every twenty-two years. There is no doubt that all stars are the seat of similar activity; even the most placid star would reveal storms if we could observe it closely enough.

Since Tycho's time, astronomers looking up at the sky no longer think of it as a vast regulated machine. "Zoo" would be a more appropriate description.

4 KEPLER, THE FIRST OF
THE MODERN MYSTICS

Kepler was one of the last professional astronomers who firmly believed (or admitted believing) in astrology. He cast an elaborate and revealing horoscope for his family, in which he says of his conception and birth: "I have investigated the matter of my conception, which took place in the year 1571, May 16, at 4:37 A.M. . . . My weakness at birth removes suspicion that my mother was already pregnant at the marriage, which was on the 15th of May. . . . Thus I was born premature, at thirty-two weeks, after 224 days, ten hours" (on December 27, 1571).

He was a sickly child, afflicted with myopia, mange, boils, a bad stomach, and a father whom he described as "vicious, inflexible, quarrelsome and doomed to a bad end. Venus and Mars increased his malice. Jupiter . . . made him a pauper but gave him a rich wife. . . . [He] treated my mother extremely ill, went finally into exile and died." Kepler was not much kinder in his description of his mother: "small, thin, swarthy, gossiping and quarrelsome, of a bad disposition." His brother was an epileptic, and Kepler himself was a hypochondriac. Yet he did not sink into self-pity. Like Job, he turned his afflictions to strength, and developed into a man with an incredible capacity for work—though he had little patience with other people.

As a child Kepler had been impressed by Tycho's comet of 1577, and he "was taken by [his] mother to a high place to look at it." He later remarked on the redness of an eclipse of the moon he had seen when he was nine years old. He began his training in a Protestant theological seminary, but before finishing he accepted a job teaching mathematics. A year later, at twenty-three, while drawing at the blackboard during a lecture he was struck by an idea that came to dominate the rest of his life: As he drew an equilateral triangle with one circle inscribed and another circumscribed he realized that the radii of the

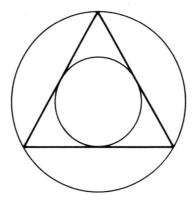

While lecturing on geometry, Kepler noticed that the ratio of the radii of circles inscribed and circumscribed about an equilateral triangle would match the orbits of Jupiter and Saturn. This set him thinking about explaining the other orbits.

two circles had nearly the same ratio as the orbits of Jupiter and Saturn. The triangle was the simplest geometrical figure, and Saturn was the outermost planet at the time. What of the remaining planets? Could they be accommodated in such a fashion? He found they could not, but he extended the quest to solid figures. To his delight he realized that there were just five regular solids whose corners would all lie on a circumscribed sphere, and whose faces would all lie tangent to an inscribed sphere. There were just five gaps among the planets—each one corresponding, no doubt, to one of the solids. With some manipulation he found the sequence which gave a tolerable facsimile of the relative distances of the planets, and he published his first book, *Cosmographic Mystery,* with an illustration of the arrangement. He felt the mystery of the universe had been solved; not only had he explained the arrangement of the solar system, he had also discovered why there were just six planets, not "twenty or a hundred." He was hypnotized by this invention and he sought more accurate planetary observations with which to refine the construction.

Tycho Brahe had a goldmine of data, and Kepler determined to "try to wrest his riches from him," because he did not believe that Tycho knew how to use the data—"as is the case with most rich people." Tycho, for his part, knew he needed help if the data were to be properly digested, so he made an overture to Kepler and they collaborated for two years, until Tycho's death. Kepler "inherited" the data and spent decades with them, attempting to determine simple planetary orbits.

Virtually every path and blind alley followed by this "sleepwalker" (Koestler's appellation) is traced in his published books. He found three simple "laws," and they lay like tiny diamonds in the sand of his texts.

Kepler found that each planet moves in an ellipse, a figure which differs from a circle and which cannot be constructed precisely, as Ptolemy and later Copernicus had attempted, by superimposing circular motions—although a close approach is possible if a large number of circles are combined. Kepler thus replaced linked circles with the simple ellipse, but he did not guess why the ellipse appeared in nature.

Kepler also showed that the times required for the planets to complete an orbit were related in a very simple way to the orbital size, and he showed that the planets move most rapidly when they are in that portion of the orbit lying closest to the sun.

In the preface to his greatest work, *The New Astronomy,* published in 1609, Kepler wrote: "What matters to me is not merely to impart to the reader what I have to say, but above all to convey to him the reasons, subterfuges, and lucky hazards which led me to my discoveries."

Kepler claimed to have been motivated by an "affection for the reader" as were Columbus and Magellan, who had provided "grand entertainment" in describing how they had gone astray on their journeys. The urge to confess had evidently been with Kepler since his youth. In his early writing, he had bared himself as did few other men of the Renaissance, and his later books fit this pattern nicely. He was constitutionally incapable of writing a pat, well-organized textbook giving only the fruits of his work.

Kepler's cosmology was an essentially theomorphic one; it centered on the trinitarian concept of God. He said:

> In the sphere [of the world] which is the image of God the Creator and the Archetype of the world there are three regions, symbols of the three persons of the Holy Trinity—the center, a symbol of the Father; the surface, of the Son; and the intermediate space, of the Holy Ghost. So, too, just as many principal parts of the world have been made—the different parts in the different regions of the sphere: the sun in the center, the sphere of the fixed stars on the surface, and lastly the planetary system in the region intermediate between the sun and the fixed stars. . . .
>
> The sun is fire, as the Pythagoreans said, or a red-hot stone or mass, as Democritus said—and the sphere of fixed stars is ice, or a crystalline sphere, comparatively speaking.

Thus Kepler retained the ancient crystalline carriers for the stars. But if the stars were confined to a sphere, in consonance with the image of God, it followed that the universe could not be infinite. Kepler, then, was faced with proving the limitation of the starry realm. He met the challenge with a turnabout on Bruno's argument that an infinite universe must be a uniform one without an identifiable center. Kepler said that when we look up at the sky we have the sensation of standing in a central void surrounded by stars. If we are in such a void the universe cannot be infinite.

The problem was to prove we are in a void, and Kepler did this in an ingenious fashion—although modern data would have led him to the contrary conclusion.

He noted that the brightest stars, which are presumably the nearest, have an angular size of about $\frac{1}{10}°$ of arc when viewed with the naked eye. They are about 10° apart on the face of the sky, so if we stand on one of them and look about we will see nearby stars whose distances are only about one hundred times as great as their diameters, as is the case for the sun in our sky. Such a sky would be a splendid display of suns—totally different from the view we have on earth. Hence the universe is not uniform and it cannot be infinite.

But Kepler's argument contained an error of fact. A few years later, Galileo found that Kepler had overestimated the apparent diameters of the stars. While the naked eye suggested that the stars covered as much as $\frac{1}{10}$°, Galileo found them to be much smaller in his telescope. In fact, he concluded that their apparent diameter in a perfect telescope might be vanishingly small.

This result destroyed Kepler's argument and dissolved another dilemma that had been introduced by the heliocentric model of Copernicus. If the earth moves about the sun, its motion should be reflected in the stars. Yet even Brahe had failed to detect an apparent yearly motion of the stars; therefore the stars must be extremely far away in comparison to the sun—in fact, they must be several thousand times farther than the sun. But if they were so far, and if they appeared to have a diameter of, say, $\frac{1}{10}$°, it followed from simple geometry that the stars must be individually several thousand times larger than the sun. Galileo's discovery that the stars appear to be mere points of light eliminated these grotesque dimensions and shrank the stars to the dimension of our sun.

Although many scientists would agree with the way history has ranked the parts of their work, this would certainly not have been the case with Kepler. The achievement he rated highest is now considered a curiosity—a relic of his ancient heritage and an example of the aberrations which occurred during the transition from medieval to modern science. Had his book *Cosmographic Mystery* been his only achievement, I think Kepler would be virtually unnoticed today, yet he found supreme joy in this realization of Greek geometrical philosophy in the structure of the solar system. To us, his configuration of nesting geometrical solids is more fitting for a garden sculpture than an astronomy book. (See plate, page 30.) It has the appearance of crank science, and absolutely no physical foundation for his architecture has come to light—nor does any scientist anticipate that one will.

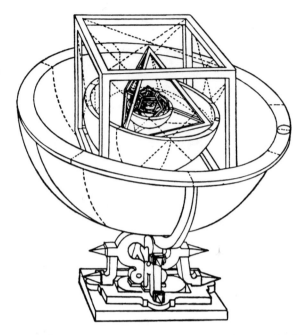

Kepler's "garden sculpture." Kepler took great pleasure in his discovery of a geometrical arrangement that would match the relative distances between the planets and the sun. A series of concentric solids supported spheres, one for each planet, at the appropriate distances. (*Reproduced, by permission, from I. B. Cohen:* The Birth of a New Physics, *Garden City, N.Y., 1960*)

This is the paradox of Kepler's work; this is what makes it so difficult to classify. We are tempted to class him with the Ancients—the Pythagoreans of Greece, for example—because he placed so much faith in the idea that the world had been cast in a mathematical mold. He felt that the world was *essentially* mathematical: by studying geometry, we study the world. This is an ancient attitude, not a modern one.

Kepler is often classed among the moderns because he was intent on finding a place for observational data in his theories: he fitted his formal constructions to the data. In this, I would concur but go on to say that even this fixation on data cannot guarantee him a place among modern scientists, because there is more to modern science than the mere satisfaction of data and the prediction of events. Modern science seeks to create a simple, yet elegant and comprehensive, description of the world—in short, an "artistic" description. The creative mind in modern science is an artistic mind, and in this sense, we have not come very far from the Ancients. But we have also developed another type: the technologist—the man of numbers and measurement, of tests and machines. This man probably existed among the Ancients as well; perhaps he was the man who constructed elaborate tables to predict the tides; he certainly built the pyramids.

It has become popular to point to the similarity of a good poem and a good physical theory; I cannot disagree with those who hold that they have much in common. Metaphor is the language of much poetry, and a mathematical law or theory is also a metaphor. No one believes that the laws of Kepler actually govern the motion of the planets, nor that they describe those motions precisely. There will always be slight deviations produced by forces too small to be of consequence, and if we wish a very precise description of the motions—for example, in the planning of an interplanetary exploration—we abandon Kepler's laws and use another set of laws, defined by Newton. Even these laws are limited; they apply only to low velocities and to distances not too many times greater than the distance to the nearest galaxy.

But there is one important difference between a poem and a theory. The theory communicates shareable knowledge about the world; the poem communicates a purely personal insight or a mood. The use of a scientific law implies a tacit agreement among scientists; they agree to be charitable toward it. The law will lose validity as soon as scientists wish it to. The validity of a poem, on the other hand, can only be judged by the poet; judgment is his personal responsibility.

The key to the paradox of Kepler's work—that he most highly valued his "garden sculpture" while we most highly value the simple laws that lie buried in his pages of arithmetic—is perhaps to be found in our own attitude. The paradox vanishes if we see the man in the light of his poetic nature, if we admit that Kepler was not a scientist but a poet, striving for personal insight and revelation rather than for shared knowledge. The fact that he dealt with planetary positions rather than the delights and pains of love, for example, may have misled us—the distinction is incidental (although it may reveal something of the man). For Kepler, the observable facts of the solar system were merely the visible signs of a transcending pattern—just as the islands of the South Seas suggest the presence of a mountainous floor beneath the water. As an oceanographer, Kepler would have imagined the undersea mountains rather than studied the islands.

When Kepler's work is set in this framework I think it is less perplexing; we may see him as the first of the modern formalists, now so acceptable among scientists. From a given set of data, he sought to construct a conceptual framework which gave meaning to the data, as a painter's conceptions give meaning to the elements of his painting, or as a novelist finds meaning in everyday events. And, like the artist or the novelist, the modern scientist ignores events and facts that do not fit his conceptual scheme; there is a popular epigram to the effect that "if the facts do not fit the theory, so much the worse for the facts."

Galileo Galilei was a contemporary of Kepler's, but his approach to experimental data was quite different from Kepler's.

Galileo never would have speculated deeply about laws that he could not verify; he wanted to know what nature had to say, not what he could imagine about nature. Galileo felt that Aristotle had given an incorrect description of the motions of rolling balls and falling weights, so he conjured up a number of alternative assumptions, computed the consequences of each assumption, and then experimented until he knew which one corresponded to the real world. The story of his dropping heavy and light balls from the Leaning Tower of Pisa may be apocryphal, but it might as well be true, because it correctly suggests that his experiments were not always highly refined: they were designed to distinguish among alternatives rather than to provide a mathematically precise and smooth curve of data.

Today, Keplers and Galileos both play their roles among modern scientists: the one type leads upward, the other recalls us to the matter of the universe.

5 GALILEO'S MESSAGE FROM THE STARS

Galileo Galilei, born in 1564, was a skilled craftsman; when he learned that a combination of magnifying lenses could make terrestrial objects appear closer, he tried several arrangements, and settled on a combination of a convex and a concave lens—the Galilean telescope was the result. In those days mirrors and lenses were widely regarded as magical devices; they were to be found at carnivals, but never in universities. There is little doubt that other men had built telescopes, but Galileo took the step that no other man appears to have taken before him: he turned his instrument on the sky. Whether he did it on a whim or because he knew the telescope could reveal unseen details of astronomical bodies, we do not know, but Galileo's description of his first nights with the telescope make up history's most exciting scientific book, *The Starry Message*, published in 1610.

The Milky Way in the constellations Sagittarius, Ophiuchus, and Scorpius. This is a three-hour exposure with a 5-inch lens, and the diameter of the field is about 21°, or slightly greater than the length of a hand held at arm's length. Note the clustering of stars, the dark markings, and the clouds of stars hanging in the background. (*Yerkes Observatory, University of Chicago*)

His excitement shows in his writing, and I can imagine how his household must have reverberated during that first season with the telescope! Within a few days he had discovered the craters on the moon, the spots moving slowly across the face of the sun, the phases of Venus, and, most significant of all, the four moons revolving about Jupiter in a tiny example of the Copernican solar system. He had incontrovertible proof of the heliocentric model—the earth moved.

His telescope fragmented the clouds of the Milky Way, and he wrote:

> I have observed the nature and the material of the Milky Way. With the aid of the telescope this has been scrutinized so directly and with such ocular certainty that all the disputes which have vexed philosophers through so many ages have been resolved, and we are at last freed from wordy debates about it. The galaxy is, in fact, nothing but a congeries of innumerable stars grouped together in clusters. Upon whatever part of it the telescope is directed, a vast crowd of stars is immediately presented to view. Many of them are rather large and quite bright, while the number of smaller ones is quite beyond calculation.

Galileo's telescopic observations carried Brahe's work to its logical conclusion: the dichotomy between the earth and the sky was dissolved; the stars were all distant suns, the planets were earths. Then as now, if a scientist claims to have made a disovery with a carnival device, he risks ridicule and his colleagues may talk about him behind his back. Just so, Kepler was asked by his Emperor, Rudolph II, what he thought of Galileo's *Starry Message*. The first copy of the book to arrive in Prague after publication in March 1610 had just reached Rudolph's hands. He immediately sent it to Kepler, who was then his Imperial Mathematician. On April 13, Kepler received a similar request from Galileo, relayed through the Tuscan Ambassador in Prague. Within a week, Kepler put a lengthy reply into the Ambassador's hands and it was sent to Galileo by diplomatic courier. A month later, in response to the varied reactions of friends, Kepler added introductory remarks and published the letter, with some alterations, under the title *Conversation with the Sidereal Messenger*.

Galileo's reasons for communicating through the Ambassador rather than writing directly are not clear, but the historian Edward Rosen suggests that Galileo may have felt remnants of embarrassment at having failed to reply to a similar request from Kepler a dozen years before, when his *Cosmographic Mystery* was published.

Kepler's introductory remarks suggest that some of his friends

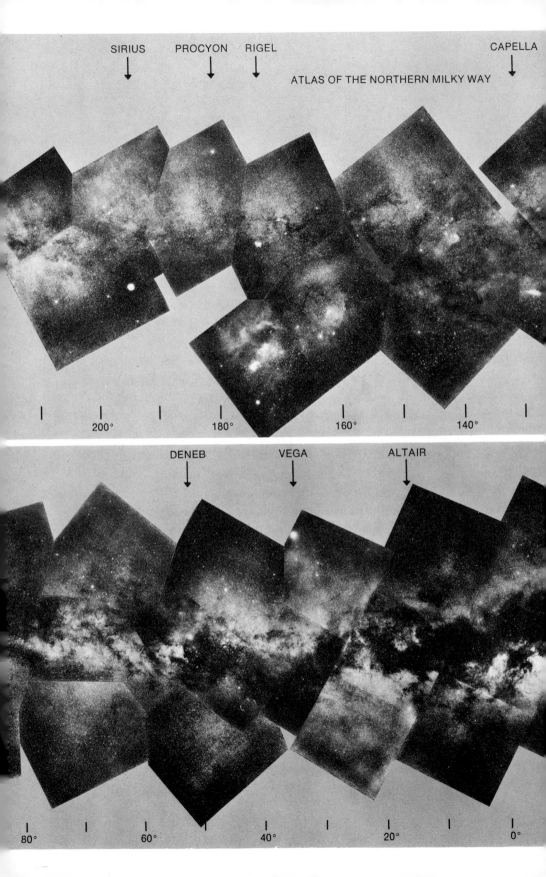

SIRIUS   PROCYON   RIGEL

CAPELLA

ATLAS OF THE NORTHERN MILKY WAY

200°   180°   160°   140°

DENEB   VEGA   ALTAIR

80°   60°   40°   20°   0°

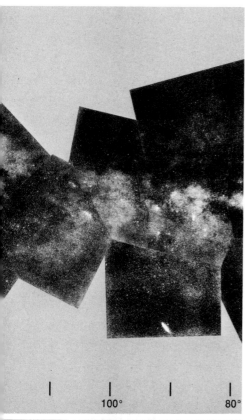

| | | | |
100°   80°

ANTARES
↓

An atlas of the Milky Way composed as a photographic mosaic by Frank E. Ross and Mary R. Calvert. Several stars are indicated. The bright star clouds of Sagittarius lie in the direction indicated by the coordinate 330°. Note the striking division of the star clouds in the vicinity of 350° to 20°. These coordinates measure galactic longitude on the old system; after the center of the galaxy was identified with the clouds of Sagittarius, the coordinates were shifted to put 0° in that direction. (*Yerkes Observatory, University of Chicago*)

ATLAS OF THE NORTHERN MILKY WAY

| | | | |
340°   320°

thought he had praised Galileo too generously, and that others thought he had criticized Galileo for making unjust claims. He wrote:

> I do not think that Galileo, an Italian, has treated me, a German, so well that in return I must flatter him with injury to the truth or to my deepest convictions.
>
> Yet let no one assume that by my readiness to agree with Galileo I propose to deprive others of their right to disagree with him. I have praised him, but all men are free to make up their own minds. What is more, I have undertaken herein to defend some of my own views also. I have done so with a conviction of their truth and with serious purpose. Yet I swear to reject them without reservation, as soon as any better informed person points out an error to me by a sound method.

The *Conversation* is a delightful treatise, dealing in turn with each part of Galileo's *Starry Message* and treating the whole matter with humor aimed "to lighten the hard work and difficulty" of the subject.

For Kepler, the most startling revelation had been Galileo's discovery of four new planets within the solar system. The bare rumor of this discovery had been relayed to Kepler by a friend, Johann Wackher, several weeks before the book arrived with the details. Kepler said that Wackher

> told me the story from his carriage in front of my house. Intense astonishment seized me as I weighed this very strange pronouncement. Our emotions were strongly aroused (because a small difference of opinion of long standing between us had unexpectedly been settled). He was so overcome with joy by the news, I with shame, both of us with laughter, that he scarcely managed to talk, and I to listen. . . .
>
> When I left Wackher's presence, I was influenced most by Galileo's prestige, achieved by the soundness of his judgment and the subtlety of his mind. Therefore I bethought myself how there could be any increase in the number of planets without harm to my *Cosmographic Mystery*, which I published thirteen years before. In that book Euclid's five solids . . . permit no more than six planets around the sun.

The Praesepe star cluster in the constellation Cancer. This photograph, taken with the 200-inch Hale telescope, shows the variety of stellar brightnesses within a cluster. There are no signs of nebulosity here.　(*Hale Observatories*)

Thus Kepler's interpretation of the structure of the solar system appeared due for destruction, because there were no more Euclidean solids with which to accommodate the additional "planets." Wackher suggested that the new planets might have been seen revolving about one of the nearby fixed stars, thus proving the plurality of the worlds. Kepler preferred to think that the new planets were merely satellites accompanying other planets as our moon accompanies us. He says, "Such was my opinion, such was his, while, our hopes aroused, we waited for Galileo's book with an extraordinary longing to read it."

When the book finally came, Kepler was "to some extent restored to life" by the news that the four new objects were moons of Jupiter. He was thus saved from having to accept the plurality of worlds—that "dreadful philosophy" that had seized his friend Wackher.

Although Kepler abhorred the idea of inhabited planets about other stars, he seemed pleased to admit the possibility of life elsewhere within the solar system. For Kepler, Jupiter's moons were evidence that the planet was inhabited. He asked: "For whose sake [are the moons] . . . if there are no people on Jupiter to behold this wonderfully varied display with their own eyes?"

His delightful imagination is revealed in the following:

It is not improbable, I must point out, that there are inhabitants not only on the moon but on Jupiter too. . . . But as soon as somebody demonstrates the art of flying, settlers from our species of man will not be lacking. Who would once have thought that the crossing of the wide ocean [of space] was calmer and safer than the narrow Adriatic Sea, Baltic Sea, or English Channel? Given ships or sails adapted to the breezes of heaven, there will be those who will not shrink from even that vast expanse. Therefore, for the sake of those who, as it were, will presently be on hand to attempt this voyage, let us establish the astronomy, Galileo, you of Jupiter, and me of the moon.

Let the foregoing pleasantries be inserted on account of the miracle of human courage, which is evident in the men of the present age especially. For the revered mysteries of sacred history are not a laughing matter for me.

Having been reassured that his *Cosmographic Mystery* had not been destroyed by Galileo's discovery—since the satellites did not need to be counted in the same geometrical structure as the major planets—Kepler next had to answer the charge that earthly astrology, of which he was a devotee, was false because astrologers had calculated the influence of the planets without knowing that Jupiter carried four unaccounted satellites. He did this easily by noting that the satellites were so close to Jupiter that their influence would be mingled with Jupiter's. With evident satisfaction he added, "In this way astrology maintains its standing. At the same time it becomes evident that these four new planets were ordained not primarily for us who live on the earth, but undoubtedly for the Jovian beings who dwell around Jupiter."

But Kepler's *Conversation* was not aimed solely to defend his own cosmology: It was principally a defense of Galileo's *Message*. Kepler had no telescope, so he could not directly confirm Galileo's discoveries. The best he could do was to point to confirmatory evidence in the writings of others and in his own books. He had known the optical principles of the telescope, but had not built one, because he felt the "minute parts of visible things at a distance are obscured and distorted"—he thought there would have been no gain. Galileo's telescope was reported to bring objects thirty-two times closer. Kepler wrote: "So powerful a telescope seems an incredible undertaking to many persons, yet it is neither impossible nor new. Nor was it recently produced by the Dutch, but many years ago it was announced by Giovanni Battista della Porta in his *Natural Magic*." Della Porta's book had indeed suggested the possibility of combining lenses in the manner Galileo later did to make distant objects clearer to the eye, and he reported showing such an instrument to friends, but Kepler says della Porta's "discussion of their construction . . . is so involved that you do not know what he is talking about."

Kepler is quite explicit in his defense of Galileo. He says, after summarizing della Porta's book, "I do not advance these suggestions for the purpose of diminishing the glory of the technical inventor, whoever he was. I am aware how great a difference there is between theoretical speculation and visual experience. . . . But here I am trying to induce the skeptical to have faith in your instrument."

In remarking on Galileo's contention that the Milky Way was merely a mass of unresolved starlight, Kepler pointed out that Brahe's suggestion that comets are formed from the nebulosity of the Milky

Way must be discarded. However, a century later this theory of comets was resurrected by the French astronomer Laplace.*

Despite what Kepler's friends may have thought of the *Conversation,* if we see it in the light of the skepticism heaped on Galileo's *Starry Message,* it is clear evidence of Kepler's remarkable sincerity and intellectual honesty. It would be difficult to see how Galileo could have read jealousy or criticism into it; he should have been immensely pleased. We have no direct evidence on his reaction to the book; Galileo was a poor correspondent. But perhaps his silence speaks for him. On several other occasions Galileo wrote biting essays in response to criticism of his claims, and there can be no doubt that he was a volatile man. If he had felt attacked by Kepler he would certainly have responded.

Galileo took an important step toward our present conception of the motion of the planets and the stars. Before him, Bruno had said that the planets were motivated like animals—they were self-propelled. Kepler discarded the concept of a living force, but he retained the idea that the planets were impelled about the sun—by the rotation of the sun, through a mechanism he did not specify. Galileo took two further steps in this direction. First he proposed universal gravitation: all bodies in the universe are attracted by all other bodies. He did not propose a specific law for the behavior of the attractive force, because he had no direct evidence and the mathematical apparatus of his day was not sufficient to elicit such evidence from the work of Kepler—that task required a Newton, and Newton was not born until the year of Galileo's death.

Second, Galileo maintained that the planets did not need to be pushed along in their orbits. Motion in free space was frictionless, he said, and it may be compared with motion of a frictionless ship on the surface of a perfectly flat sea. With this imaginary ship, Galileo performed a series of "thought experiments." He imagined that the smooth ship on the flat sea would glide "continually around our globe . . . if

* Another remark of Kepler's, that Galileo had revealed the true character "of the Milky Way, the nebulae, and the nebulous spirals [nebulosis convolutionibus]," is enigmatic to me. As far as I know, this is the first use of the word "spiral." Kepler could not have seen nebulous spirals himself as he had not yet used a telescope, and Galileo did not use the word in his *Starry Message.* John Herschel, in the nineteenth century, is the next astronomer to have used the description, and only in the middle of that century was the spiral nature of many nebulae generally acknowledged.

all external impediments could be removed." In this way, he dissected phenomena as a biologist might separate the organs of an animal. This type of thought experiment has proved to be one of the most marvelous inventions of man—it led Einstein to his theory of relativity, and it led Newton to his greatest discoveries.

Yet even Galileo, who had opened the Milky Way with his telescope, could not totally free his mind from its earthly modes. His description of unimpeded motion was the peninsula from which he reached toward his successor, Newton, but it was the site of an ancient temple: the concept that circular motion was natural.

Galileo was primarily an earthly physicist, and his ideal of unimpeded motion was along the spherical surface of the earth. He wrote:

> For I seem to have observed that physical bodies have physical inclination to some motion (as heavy bodies downward). . . . And to another motion they have a repugnance (as the same heavy bodies to motion upward). . . .
>
> Finally, to some movements they are indifferent, as are these same heavy bodies to horizontal motion, to which they have neither inclination (since it is not toward the center of the earth) nor repugnance (since it does not carry them away from that center). And therefore all external impediments removed, a heavy body on a spherical surface concentric with the earth will be indifferent to rest and to movements toward any part of the horizon. And it will maintain itself in that state in which it has once been placed; that is, if placed in a state of rest, it will conserve that; and if placed in movement toward the west (for example) it will maintain itself in that movement. Thus a ship, for instance, having once received some impetus through the tranquil sea, would move continually around our globe without ever stopping . . . if all external impediments could be removed.

Galileo was almost correct in this description of motion on the surface of the earth. Only a very precise measurement would have disclosed his error, and yet the measurement would have required him to describe the motion in totally different terms. Because the earth rotates, an artillery projectile curves toward the right when fired in the Northern Hemisphere. Although Galileo was clearly aware of the rotation of the earth, he had no experience with long-range artillery and his description of motion would not have explained the behavior of the projectile, nor could he have understood the elliptical planetary orbits

Kepler had discovered. Galileo's mind was bound to circular motions concentric with the earth and with the sun.

The possibility that natural unimpeded motion would follow a straight line was rejected by Galileo on metaphysical grounds, with the comment: "Straight motion being by nature infinite (because a straight line is infinite and indeterminate), it is impossible that anything should by nature have the principle of moving in a straight line; or, in other words, toward a place where it is impossible to arrive, there being no finite end."

Shortly afterward, the philosopher Descartes rejected Galileo's argument and adopted constant velocity along a straight line as the ideal of unimpeded motion. In effect, he showed that Galileo had assumed too much and had failed to make a crucial distinction among the forces acting on an object such as a ship.

Imagine yourself on a bluff overlooking the sea on a calm day; a ship steams past and heads toward the horizon. Descartes asks you to imagine that the ship would fly off into space along a straight line if it were not for the competition between the buoyancy of the water, the downward pull of gravity, the forward thrust of the propeller, and the backward thrust of the water resistance. If this sounds like thinking of a person as a collection of bones, flesh, hair, etc., instead of a personality, we can understand why Galileo did not make the step. His experience did not require it, and it would have seemed unnatural to him. But these distinctions provided insight into a whole new field of phenomena.

Christiaan Huygens, born in 1629, bridged the span between the death of Galileo in 1642 and Newton's first publication of his *Principia Mathematica* in 1687. Huygens perfected the pendulum clock and the telescope, legacies from Galileo, and went on to develop the wave theory of light—a radically new idea.

Huygens performed the first calculation of the distance to a star by comparing its apparent brightness to the apparent brightness of the sun, and by assuming that the star would be as bright as the sun if it were as close as the sun. For the brightest star in the sky, Sirius, he found a distance two hundred thousand times the earth's distance from the sun, a value which has been increased by a factor of 20 in modern times. Huygens's assumption of equal brightness underestimated the brightness of Sirius by a factor of 400.

Huygens adopted Descartes' idea that the planets would fly off into space except for the force of the sun, which bends their orbits into closed curves. Assuming that the orbits were perfect circles (he knew they were not, but he could not analyze elliptical motion), he was able to show that the gravitational force of the sun weakens as the square of the distance from the sun.

There the matter stood when Newton appeared: The picture was fragmentary and incomplete, but elements of a design had begun to emerge.

6 NEWTON

To my mind, the most brilliantly conceived diagram in the history of science is Isaac Newton's imaginary view of the earth with a single tall mountain. (See plate, page 48.) On the top of the mountain we see a cannon firing a series of projectiles. The first falls near the foot of the mountain; the second is fired more swiftly and it circles part way around the globe before falling to the earth. Finally, a projectile is fired with precisely the right velocity; as it falls toward the earth, the surface of the earth curves away underneath and the projectile stays at a constant height above the ground while circling the entire globe—it has gone into orbit. With this simple diagram, Newton carried astronomical dynamics from Galileo to the day of artificial earth satellites.

A single addition to the diagram would carry us to interstellar space and open the way to the exploration of the Milky Way: a fourth

Isaac Newton (1642–1727). His invention of the concept of universal
gravitation and his formulation of the laws governing the motions of material bodies laid the groundwork for the modern astronomer's view of
the Milky Way as a rotating system of stars. No other man, with the
possible exception of Charles Darwin, has had a greater influence than
Newton on our present view of the universe and man's role in nature. As
this portrait hints, Newton was an arrogant, irascible, and lonely man;
his planets were lonely objects drifting through space under the control
of the sun. (*Yerkes Observatory, University of Chicago*)

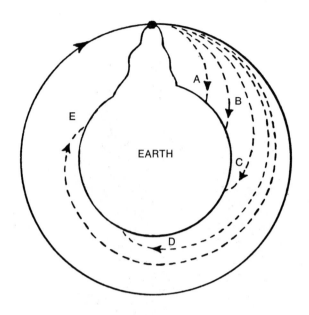

From cannon ball to artificial satellite. This diagram, taken from a popular book by Newton, illustrates the power of Newton's concept of motion; it also comprises the first explicit prediction that man would one day launch an artificial satellite into an orbit about the earth. A cannon on a mountain top may fire a ball to point A; with more power, the ball may fly farther around the world, to B, or C, or even E, before falling to the ground. Given just the right impulse, the cannon ball could fall toward the ground at just the proper rate to pass entirely around the planet; it would then be in orbit. With a sufficient impulse, the ball would escape from the earth's gravitational field altogether. (*Reproduced, by permission, from Stephen Toulmin and June Goodfield:* The Fabric of the Heavens, *New York, 1961*)

projectile may be shot so swiftly that it leaves the earth altogether, flying off in a trajectory that is only slightly curved by the pull of our globe. This projectile goes into orbit about the sun, or if its velocity is greater still, it leaves the solar system and is lost among the stars.

This fourth projectile would illustrate the idea of Descartes and Huygens that a body which has been pushed will move forever in a straight line with a constant velocity, unless it is pushed again or otherwise deflected by an external force. If I drop a pencil in an airplane, or if I drop a stone while riding the earth through space at 18 miles per second, I need not wonder why the pencil or the stone stays with me and appears to drop straight toward the ground. When I let go, the pencil was moving along with my hand and with the chairs and the floor of the airplane, and this forward motion is not lost while the pencil falls. The forward motion of the airplane is unnoticed because it is constant. Changes in the motion of the pencil require explanation (the downward acceleration due to gravity in this case), but the motion itself does not.

From such structural elements, provided for the most part by his predecessors, Newton synthesized a new physics. He performed no crucial experiments in mechanics nor any important observations of the planets, but he imbedded the knowledge of his day into a simple yet comprehensive framework. The two principal components of his physics dealt with the nature of motion and the idea that all matter attracts matter.

Before Newton and Galileo, motion had been regarded as an activity requiring an agent. Aristotle had pointed out that two horses can move a wagon twice as fast as one horse, and if the horses die the wagon will stop because the horses died. The logical extension of these facts seemed to be that all changes of position will occur more rapidly if a greater effort is expended. But according to Newton, the wagon stops not because the horses died, but because the friction of the prostrate bodies halted the motion. Further, although a greater effort is needed to keep a wagon moving more rapidly, planets do not behave as wagons. Newton said we must visualize planets and stars as completely isolated from the friction of wheels or the drag of a fluid in space. In his eyes, the stars and planets became a society of isolated individuals floating freely through empty space, attracting each other by a force that could be described precisely but whose origin and whose agent were incomprehensible. There is a striking parallel between this view of the universe of stars and what we imagine to have been Newton's view of the society of men. To emphasize this parallel might be to dig too deeply for a psychological interpretation of Newton's physics, but to ignore it would be to ignore Newton's sense of isolation from his contemporaries and to underestimate the radical change he brought to cosmology.

Like many great men, Newton saw himself as a divinely inspired prophet revealing the glory of the heavens to man so that man might know the wonder of God. He never knew his father, who died shortly before his birth, and his mother soon remarried, leaving him with his grandmother. As a child he played alone, building mechanical toys, carts, water clocks; and from this experience he developed an intuitive sense of motion and force.

When Newton reached the age of twenty he confessed a craving to kill his stepfather; but when civil war broke out in England he returned to the countryside to live with his mother and became quite close to her.

Those few years were the "miraculous years" of his creativity—the most highly creative in the history of science—when ideas for the analysis of light and color, the basis of calculus, and his concept of motion and universal gravitation sprouted in Newton's mind. He was highly devout, and as a young man he knew as much about the Bible as do most theologians. For him the Bible was the revelation of the Word of God, and, by extension, the book of nature was a revelation of God's works. He felt he had been chosen to reveal these works to man; science became a form of worship for him. Mathematics was a sharp steel tool in his hands, cutting into the dilemma and uncertainty of the world. A recent biographer, Frank E. Manuel, says, "The discovery of his mathematical genius was his salvation." Planets obeying his mathematical laws were the bedrock of his certainty, and the creations of his mind were the flowers of his personal "garden."

According to Manuel, the young Newton assimilated God with the earthly father he had never known, and whose favor he had never enjoyed. Through his creative years, he was "sustained by a consciousness of the direct personal relationship between himself and God his Father, uninterrupted by a mediator." Newton, in other words, thought he was a Christ.

Though his work was inspired by religious sentiment, he was one of the first to proclaim the religious neutrality of science. He claimed: "We are not to introduce divine revelations into philosophy, nor philosophical opinions into religion." But to the modern view, he did not altogether succeed in keeping distinct his religion and his cosmology. He retained a role for the Deity in his universe; when he stood at the edge of his knowledge, Newton called upon the "counsel and contrivance of a voluntary Agent" who had performed the act of creation and had set the universe in motion. Newton and most scientists of his day were teleologists. They saw divine purpose in every detail of the world, "in the structure of an animal molar, in the hair of a cheese mite, in the orderly movement of the planets," according to Manuel. Of the solar system, Newton wrote:

> The six primary planets are revolved about the sun in circles concentric with the sun, and with motions directed toward the same parts and almost in the same plane. Ten moons are revolved about the earth, Jupiter, and Saturn, in circles concentric with them, with the same direction of motion, and nearly in the planes of the orbits of these planets; but it is not to be conceived that mere mechanical causes could give birth to so many regular motions, since the comets range over all parts of the heavens in very eccentric orbits. . . .
>
> This most beautiful system of the sun, planets, and comets could only proceed from the counsel and dominion of an intelligent and powerful Being. And if the fixed stars are the centers of other like systems, these, being formed by the like wise counsel, must be all subject to the dominion of One.

This is the standard argument from the orderliness of the universe to its divine construction. Newton then closes the logical circle of the argument by saying that God had placed the stars at immense distances from each other, lest by their gravity they should fall on each other. Like all arguments for the existence of God, this one convinces only those who are prepared to believe—as Newton certainly was. It fails in the face of skepticism, as the Frenchman Laplace later proved when he developed a descriptive theory for the origin of the solar system which explained the orderliness as a natural consequence of Newton's own physical principles. Newton had been entirely correct in saying that the present state of the solar system cannot be explained in terms of forces *presently* acting, but Laplace showed that there was an alternative to calling in the Deity—he suggested that the solar system had once been a nebulous swirl of matter that flattened while it cooled and contracted, thus leaving the planets in circular orbits aligned nearly in the same plane.

Newton was among the first to speculate on the evolution of the heavens, and this was well before evolution of terrestrial forms or animals was seriously contemplated. To a friend, the Reverend Richard

Bentley, he wrote a number of letters discussing his cosmology, and one contains the following description of the birth of stars from a uniformly distributed stratum:

> It seems to me that if the matter of our sun and planets and all the matter of the universe were evenly scattered throughout all the heavens, and every particle had an innate gravity toward all the rest, and the whole space throughout which this matter was scattered was but finite, the matter on the outside [i.e., in outer parts] of this space would, by its gravity, tend toward all the matter on the inside and, by consequence, fall down into the middle of the whole space and there compose one great spherical mass. But if the matter was evenly disposed throughout an infinite space, it could never convene into one mass; but some of it would convene into one mass and some into another, so as to make an infinite number of great masses, scattered at great distances from one to another throughout all that infinite space. And thus might the sun and the fixed stars be formed, supposing the matter were of a lucid nature. But how the matter should divide itself into two sorts, and that part of it which is fit to compose a shining body whilst all they continue opaque, or all they be changed into opaque ones whilst he remains unchanged, I do not think explicable by mere natural causes, but am forced to ascribe it to the counsel and contrivance of a voluntary Agent.

This paragraph became a prophecy of astronomical thought for the two centuries following Newton. When he imagined this evolution of matter from a uniform stratum into individual spheres—some shining as stars, others opaque as planets—he had no direct evidence. This image was born from his concept of universal gravitation.

The remaining story of cosmology—at least for the purpose of this book—has been the struggle to capture the evidence for Newton's cosmology and weave it into a convincing picture of the heavens, describing the actual objects into which matter had convened, and attempting to define the next set of questions: Into what state is the universe now evolving? What are the processes by which our stellar universe has been molded and will be molded?

7 THE NEBULOUS PATCHES

Orion is a large, rectangular constellation of the winter sky. Three moderately bright stars compose a "belt," and from the belt hangs a "sword" of three faint stars. The central star of the sword appears strangely soft to the naked eye; it is a star, but it is not quite a star.

Andromeda is a constellation of the autumn sky; in its center is a large, nebulous patch which is quite easily visible on a dark night. It is clearly not a star; its image is spread into a long oval and it contains no sharp point of light. This patch, the brightest nebula in the sky, appears on star charts predating Galileo, but it was Simon Marius, a contemporary of Galileo, who gave the first adequate description when he wrote that it resembled "the light of a candle which one sees from a distance in the night through a piece of transparent horn." The plate on page 55 shows the nebula in Andromeda as it appears in a pair of powerful binoculars.

There are several other patches visible to the naked eye. In the constellation Hercules lies a perfectly round patch, shaded smoothly toward the center and slightly smaller than the Andromeda nebula.

These bright patches appear to bear no relation to the Milky Way; they are strewn at random over the sky. But along the Milky Way there also lie numerous smaller condensations of light that the eye does not quite accept for stars. The Milky Way itself is knotted and ragged; its edge has the appearance of an old flag.

When Galileo found that the Milky Way was fragmented into stars by his telescope, he was delighted to have at last "solved" the mystery of that band of star light. In his book *The Starry Message,* he says:

> But it is not only in the Milky Way that whitish clouds are seen; several patches of similar aspect shine with faint light here and there throughout the aether, and if the telescope is turned upon any of these it confronts us with a tight mass of stars. And what is even more remarkable, the stars which have been called "nebulous" by every astronomer up to this time turn out to be a group of very small stars arranged in a wonderful manner. Although each star separately escapes our sight on account of its smallness or the immense distance from us, the mingling of their rays gives rise to that gleam which was formerly believed to be some denser part of the aether that was capable of reflecting rays from stars or from the sun.

Galileo was not a cataloguer. He made no lists or detailed descriptions of the Milky Way nebulosities so we cannot know which ones he observed, but he was clearly convinced that most, if not all, of the nebulosities would be resolved into clusters of faint stars. His telescope has recently been re-examined and actually used at night by a com-

---

The Andromeda Nebula. This oval patch of light, nearly the size of the face of the moon, was apparently the first "nebulous star" discovered. In a pair of binoculars, it appears as a faint, ill-defined glow surrounding a central nucleus. Note the small companion toward the upper-left corner from the principal nebula. The ring below is a photographic effect produced by the reflection of starlight from the back of the glass photographic plate. (*Harvard College Observatory*)

patriot of his, Giorgio Abetti, and it is quite reasonable that Galileo attributed most of the nebulosity to the blur within his instrument. Abetti says:

> Some years ago at the Observatory of Arcetri observations were made with the first telescopes of Galileo (with more ease and comfort than Galileo could have had). The telescopes were attached to a larger companion telescope equipped with an equatorial mount and a clock-work mechanism, so that they followed automatically the diurnal motion of the celestial sphere. . . . Through these telescopes we were able to see the same bodies observed by him: the sun, the moon, Jupiter, and Saturn. We were able to determine the accuracy of his observations and the optical imperfections which these instruments necessarily possessed, as well as to admire the sharpness of his eyesight and his intuition. His first object lens, which is larger than the others, seemed to us the best of all. Although it is now broken in several pieces, it shows better optical characteristics and a resolving power of about 10″, which therefore enabled Galileo to separate the disk of Jupiter from its satellites up to this angular distance.

Galileo thought he had banished true nebulosity from the sky, leaving it populated only with stars, but he went too far. His successors found many patches that would not disintegrate into stars, even in vastly improved telescopes. The most spectacular of these was the peculiar "star" in the center of Orion's sword. In the telescope the star became a pale-blue mist surrounding a tight group of four stars, as shown on page 57. Huygens wrote that the stars in the center of this nebula "seemed to shine through a fog in such a way that the surrounding space . . . appeared much brighter than the rest of the sky, which was very placid and extremely dark, and seemed to have a hole through which a more lucid region could be perceived." William Derham, an English clergyman and dabbling astronomer, suggested that this patch of light was a visible entrance to the empyrean.

By 1700, about ten nebulae were known, but no systematic study had yet been attempted. The nebulosities were treated as mere curiosities.

The central region of the Orion Nebula showing four stars arranged in a small trapezium. These stars are responsible for most of the nebular light. (*Lick Observatory, University of California*)

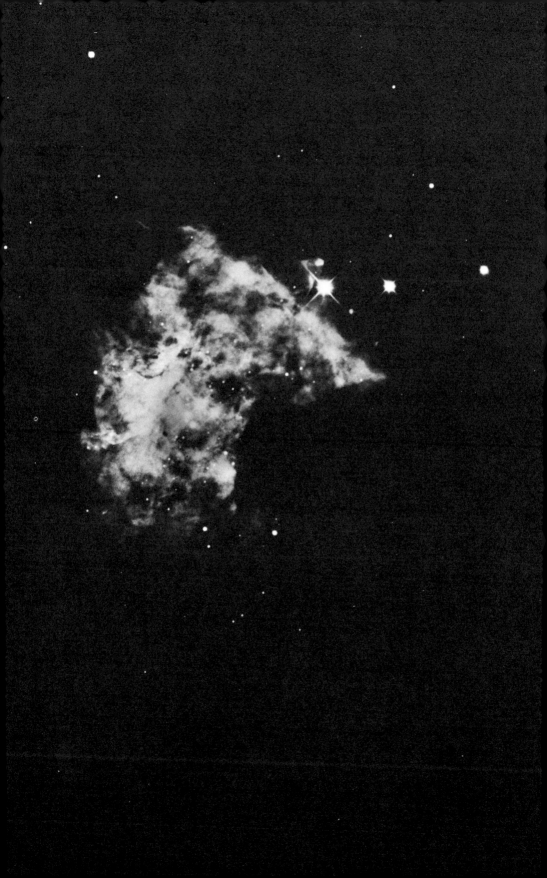

The Abbé Nicolas Louis de La Caille published the first extensive list of nebulous stars in 1755. In common with many astronomers of the eighteenth century, La Caille had been trained originally for the priesthood, but this education was easily diverted toward mathematical and astronomical studies. On a voyage to the Southern Hemisphere, La Caille measured the distance to the moon and examined the sky quite thoroughly, collecting data for an immense star catalogue. La Caille's description of the classification scheme implies his belief in the existence of nebulosity. He wrote in the *Memoires of the Royal Academy of Paris:*

> The first [class] is nothing but a vaguely terminated whitish space, more or less luminous and frequently of a very irregular form; these spots ordinarily resemble quite well the nuclei of faint comets without tails. The second species of nebulae comprises stars which are only nebulae in appearance and to the unaided eye, but which one sees at the telescope as a cluster of distinct stars, quite close together. The third species is that of stars which are really accompanied or surrounded by white spots or by nebulae of the first species.

But reading further we find that La Caille after all had retained the Galilean view that nebulosity was an optical effect of unresolved stars. He said, in conclusion, "Perhaps it would not be too hazardous to advance [the idea] that the nebulae of the first species are only like small portions of the Milky Way, spread in different parts of the sky, and that the nebulae of the third species are only stars which, relative to us, are placed in the straight line along which we view these luminous patches."

The first substantial evidence on the nature of the nebulous patches came from Jean-Jacques Dortous de Mairan, whose book titled *Physical and Historical Treatise on the Aurora Borealis* (1733) must stand as one of the most remarkable scientific works of the eighteenth century. This was the first application of geophysical data to an astronomical problem; Mairan suggested that the earth and the sun were connected by bridges of gas, and he made the first modern analysis of the nature of stars.

Mairan's starting point was an extensive study of the geographical distribution of Northern Lights. Before his time these colored lights were generally attributed to volcanic exhalations, but Mairan showed that the great height at which they occurred and their restriction to the far north argued against the volcanic hypothesis. Why, he asked, wouldn't the phosphorescent vapors be seen concentrated in the neighborhood of volcanoes?

He proposed that the Northern Lights were caused by wisps of the solar atmosphere ejected from the sun and descending into the earth's atmosphere.

During a total solar eclipse, when the sun's face is obliterated by the body of the moon and the sky becomes dark, a silvery glow surrounds the sun and extends tendrils to a distance of several times the sun's diameter. Mairan suggested that this "corona" of the sun is evidence of gas streaming outwards, and that this gas reaches the earth, where chemical reactions in the atmosphere cause the emission of light. In favor of this idea, Mairan could only offer a series of indirect arguments. (His hypothesis received final proof recently when space probes sent into orbit about the sun detected the streams of gas—flowing outward among the planets. The solar "wind," it is now called.)

Mairan went further in his speculations, and suggested that streams of gas emanating from other stars might appear as nebulous patches in the sky. He noted that the sun is probably a typical star and that its exhalations would not be visible from even the nearest star. But "monstrous" examples surrounding other stars might be visible. "After all," he said, "despite the admirable uniformity which reigns in the operation of nature, the universe has monsters in the large as well as in the small." Wasn't it the purpose of comparative astronomy, as of anatomy, to investigate such monsters?

Mairan's book must have been fairly widely read; it was reprinted in 1734 and 1735, and a second edition was published in 1754. His own conclusion was unequivocal: truly nebulous matter exists in the sky.

In 1705, the Englishman Edmund Halley opened a new branch of astronomy with the publication of his book *A Synopsis of the Astronomy of Comets*. Important new studies of nebulous patches resulted from the new interest in comets.

Halley analyzed all available observations of comets. He claimed that one series of particularly bright apparitions—in the years 1531, 1607, and 1682—was to be explained by a single comet in a seventy-six-year elliptical orbit about the sun. He thus contended that this object was a member of the sun's family, not a mere will-o'-the-wisp. Halley calculated that this comet would reappear near Christmas 1758, and knowing that he would not be alive (he would have been 102 years old) he made a special plea that astronomers look for the comet and remember that it was an Englishman who had first predicted its return. The world indeed remembered, and the comet was given his name when it returned. Its most recent return was 1910. Historical records have revealed several earlier trips near the sun, and the Bayeux tapestry commemorating the Battle of Hastings shows the comet on its return in 1066.

Once Halley had shown that comets could be successfully analyzed by Newtonian physics, comet hunting became a popular sport and a frequent source of quick fame among astronomers—professionals and amateurs alike.

An ardent comet hunter would spend evenings and mornings sweeping the sky with a telescope, looking for a faint patch of light, softer than the image of a star and not unlike many of the fixed nebulosities in the sky. Finding one, he would mark it on a chart and return to that part of the sky the next night, hoping to find the patch had moved. If it had not moved he could assume that it was a "nebulous star," and he would make a permanent notation on his chart to avoid confusion the next time. If it had moved he would wait for more opportunities to measure the position, and when he had obtained at least three good observations he would compute the orbit, or simply announce the observations, depending on his mathematical zeal.

Thus comets were added to eclipses and planetary motions in the kit bag of the observing astronomer, but the study of nebulosities remained quite casual, and virtually all of the discoveries before 1790 were considered a byproduct of the search for comets.

In 1784, Charles Messier and Pierre Méchain listed 103 nebulous objects and clusters that they had marked on their charts in the course of three decades of comet hunting. This list was the starting point for later

The nebula M87 surrounded by a host of fainter nebulae. This is No. 87 in the list of Charles Messier; he described it as being without stars. To this day, no stars have been discovered, but the faint images surrounding it are thought to be globular clusters, like that shown in the plate on page 62. (*Lick Observatory, University of California*)

searches that opened the Milky Way and that led astronomers to external galaxies. A more modest publication of greater ultimate significance is difficult to imagine, but Messier and Méchain totally refrained from comment; they simply listed the positions of the nebulae and clusters, and gave very brief descriptions with which an observer of moderate experience could verify his identification if he found a fuzzy patch of light in his own telescope.

Charles Messier was born in 1730. Before he was twelve his father died, and an older brother acted as guardian. At the age of eighteen he witnessed a marvelous eclipse of the sun and his mind turned to astronomy. At twenty, with little training but with fine penmanship and some practice in drawing, he went to Paris, where he was taken as apprentice to Delisle, a prominent astronomer who set him to keeping accounts of observations. Messier's condition was meager, and he had little opportunity for study, but he soon engrossed himself in night work at the telescope, the "kind of work which suited him best," according to his memoires.

In 1759 he suffered a painful disappointment in failing to be the first to detect Halley's comet on its predicted return to the sun—an honor which he and the world fully expected him to achieve. The orbit was uncertain, but Messier, relying too heavily on the map which had been provided, spent eighteen months peering through his telescope in fruitless search near the predicted positions. On Christmas of 1758, a peasant found the comet with his naked eye and during the following weeks several others succeeded. Word of these discoveries failed to reach Paris, and a full month passed before Messier discovered the comet himself. At Delisle's insistence, the discovery was kept secret until another month had passed and the reappearance of the comet had become common knowledge. When at last Delisle permitted Messier to

---

The globular star cluster M3. This cluster, nearly spherical and containing hundreds of thousands of stars, is one of several hundred similar members of the Milky Way. The cluster is held together by gravitational attraction among the stars, which pursue orbits through the swarm. The stars are actually quite far from each other, and no collisions are expected, despite the optical effect which makes them look as though they were in contact. (*120-inch photograph, Lick Observatory, University of California*)

announce his independent discovery, Messier was received with derision and accused of falsifying his observations.

This was not the only time Delisle prevented Messier's announcements of cometary observations, and his behavior is all the more odd because neither man cared a whit about calculating the orbits. Shortly afterward, Delisle abandoned science for "devotional practices," leaving Messier with the freedom he must have craved.

For the remainder of his life Messier chased comets—he was the "ferret of comets," in the words of Louis XV. A story is told of him which, true or not, must have had some sound of truth in it or his contemporaries would not have repeated it. The final illness and death of his wife prevented Messier from the discovery of an anticipated comet, a feat performed by a contemporary, Montagne de Limoges. When a friend spoke to him of his loss, he replied, "Alas, I had discovered a dozen comets, and Montagne had to take away from me the thirteenth." Realizing his faux pas, he then said, "Ah, that poor woman!" but he was reportedly still thinking about the comet.

The detachment with which Messier described the nebulosities and clusters perhaps reflected his preoccupation with comets. Nowhere in the discussion of the catalogue is there a hint of systematic classification of the nebulae, nor is there a discussion of the problem of distinguishing true nebulosity from unresolved groups of stars. To Messier and Méchain, these questions were sidetracks from their pursuit of comets. For other men, the nebulosities became hints to the structure of the universe.

M16, a nebula with stars. Looking at this modern photograph, there can be little doubt that this nebula is composed of gas rather than being a cloud of stars. Note the mixture of bright and dark features, the many dark globules, and the bright rims. (*Lick Observatory, University of California*)

PART II
A PROBLEM OF DEFINITION

8 EARLY SPECULATIONS ON
THE ORIGIN
OF THE SOLAR SYSTEM

The spectacular form of the system of planets—its arrangement in aligned circular orbits—is an enticement to speculation on origins.

Early speculation on the origin of the solar system led outward to the stars and nebulae; it provided a link between nebulae and the solar system.

Newton refused to be seduced into such mind-play; he said: "I frame no hypotheses; for whatever is not deduced from the phenomena is to be called an hypothesis; and hypotheses, whether metaphysical or physical, whether of occult qualities or mechanical, have no place in [physics]." Newton did not try to "explain" gravity, for example, because he knew of no phenomenon among the planets that required a more complete description of gravity than his simple formulae had given. For him, explanation was equivalent to mathematical description; "meaning" had no meaning beyond mathematical consistency, and this attitude fostered what later became "science" as opposed to "mysticism," or the revelation of meaning behind events. Newton's satisfaction with mathematical description is not far removed from the Pythagorean tradition of the ancient Greeks; it is in fact a very strong link between the ancient and the modern philosophers, and it emphasizes the historical continuity of man's attitude toward his environment.

The most famous of Newton's immediate predecessors, the French philosopher René Descartes, was born in 1596. He had a fine education in science and he took great delight in the methodical use of his mind to explore the universe. His book *Discourse on Method* proclaimed the unlimited power of pure intuition—so long as it started from absolute certainty and proceeded in small and certain steps. According to Descartes, certainty resided in the phrase "I think, therefore I am."

Descartes became dissatisfied with Kepler's theory of planetary motion because it required the sun to act on the planets across empty space. He thought this was bad physics, and he suggested that the planets are captive in an ethereal vortex which swirls around the sun, carrying them along as water carries bits of wood in a whirlpool. For almost a century this concept of vortices was bruited about in all corners of Europe and Scandinavia. The world was grateful to the little Frenchman for being, and thinking, and therefore being able to construct such a compelling model for the operation of that great machine, the solar system. Uncertainties were whirled away in the thoughtful eddies of their speculations.

Newton preferred his own idea of a central force, so he set out to sink Descartes. He invented mathematical techniques for studying fluid flow and investigated the properties of vortices. It is not clear just why he thought it fair to place one of *his* vortices at the center of Descartes' solar system, but he did, and the result was disastrous for Descartes' theory. No planet would stay in its orbit until Newton assigned to the vortex a density equal to that of the planet—and as each planet had a different density, and implied a different density for the vortex, Newton could only shrug his mathematical shoulders.

Kepler's laws fared no better; they required a preposterous vortex, and Newton was moved to say: "The hypothesis of vortices is pressed with many difficulties. That every planet by a radius drawn to the sun may describe areas proportional to the times of description, the periodic times of the several parts of the vortices should observe the duplicate proportion of their distances from the sun; but . . ." and so on. Every phenomenon seemed to require a new vortex, and Descartes' theory began to look silly, although it was given up only with great reluctance. For half a century scientists were divided into two camps: Newtonians and Cartesians—those who were willing to accept mathematical formulae and hold speculation in abeyance, and those who insisted on finding underlying causes even when the details of their speculations did not quite hold together.

There appears to be a close parallel between Newton's philosophy—no unessential hypotheses—and modern-day existentialism. Newton existed alone in his own universe; he sought solace through science, and he found it through mathematical abstraction. For him "meaning" was meaningless, and he reminds one of Jean-Paul Sartre or Samuel Beckett in this respect: we cannot communicate meaning to one another; the best we can do is describe events—and that rather sparsely. Descartes, on the other hand, was essentially an optimist. That is, he took a "religious" approach to the world: we are not alone; the world does have meaning, and the meaning is in our own rationalizations. We can find the unity permeating the universe *if* we will use our minds rationally. Descartes focused on the mind as a tool for enlightenment, and he opened the door on centuries of debate about the meaning of "rational"—a debate that still stands at the center of philosophy.

Emanuel Swedenborg, born in 1688, the year of Newton's peak creativity, was a Cartesian, and he appears to have been the first to speculate on the origin of the planets after Copernicus had placed the sun at the center of the system.

This man combined the creativities of the rational mind and of the visionary. It is true that he could be called either a "mystical scientist" or a "rational mystic," but it would be truer of him to say that he was occasionally a rationalist and more often a mystical seer.

Historically, Swedenborg stands at the point of the initial separation of natural philosophy, which became "science," and religion. As the separation proceeded through the eighteenth and nineteenth centuries, it became popular to think that science was an activity of appalling objectivity, while artistic creation was an activity of enlightened, but self-indulgent, subjectivity. In a diluted form this attitude persists today, and although artists and scientists would deny its validity, there is no denying that the work of men like Swedenborg, who seemed to live *two* lives, gives it some credence.

Swedenborg's father was a clergyman of the Swedish Lutheran Church. The boy graduated from Uppsala University in 1709, spent six years visiting and studying in Europe and England, and then became Assessor Extraordinary in the Swedish College of Mines. For the next twenty-five years he undertook a wide-ranging series of experiments in astronomy, economics, chemistry, metallurgy, and mechanics; he also wrote the first Swedish book on algebra. In 1743 he began having psychic experiences, and they continued until his death at the age of 84. He did not marry, but he admitted that he "was always partial to the company of ladies."

In his later years, Swedenborg often reported having communicated with the spirits of the departed wives and husbands of persons still living. He brought back news of life in the hereafter—for the most part good news. On one occasion, however, an archbishop, whose partner in three-handed cards had just died, met Swedenborg and asked him, "By the way, Assessor, tell us something about the spirit world. How does my friend Broman spend his time there?" Swedenborg answered, "I saw him but a few hours ago shuffling his cards in the company of the Evil One, and he was only waiting for your worship to make up a [threesome]."

In one of his most delightful books, Swedenborg describes conversations with angels from other planets. This book might well be classed with the great fantasies of Western literature, such as *Gulliver's Travels,* but it has not been, presumably because Swedenborg did not have a reputation as a writer of fantasy. He has been taken at his word, and his insistence that the conversations were real—even on his deathbed—may have prevented their being accepted as literature. But they are rich with insights, and I prefer not to be distracted by questions of his motives, his occult powers, or his possible delusions.

In one passage, Swedenborg describes his encounter with Aristotle, and it is tempting to suppose that the passage is actually a self-portrait of Swedenborg:

> He evolved from his own thought the things he had written, and thence produced his philosophy; so that the terms which he invented

and imposed on subjects of thought, were forms of expression by which he described interior things; also that he was excited to such pursuits by the enjoyment of affection, and by the desire of knowing the things of the thought and intellect; and that he followed obediently what his spirit had dictated.

This seems to me a fair description of some modern scientists.

Swedenborg's attitude toward science is suggested in his description of a conversation with the angels of the planet Jupiter. He says that they were very skeptical of science, about which "they know nothing whatever, nor do they wish to know," because they had heard that scientists on our earth boasted of having become wise through science, whereas they had only acquired "things that are of mere memory." But Swedenborg reports that he protested this opinion and that he said to the angels: "On this earth the sciences are means of opening the intellectual sight, which sight is in the light of heaven." Thus, experimentation and rationalization were for Swedenborg a pathway to the problems of the spirit and the meaning of existence.

The underlying principle of Swedenborg's philosophy of nature was the idea that the structures of the smallest and the largest are similar—all matter conforms to ethereal swirls: atoms are magnetic vortices; the solar system is propelled by a vortex; the Milky Way is formed of a rotating mass of stars; etc. He saw the spiritual world coordinated in the Greatest Man: a super-spirit constructed in the form of a man and composed of spirits from all of the planets, each planet contributing a particular attribute.

These ideas are strikingly similar to "likepartedness," a concept which the ancient Roman Lucretius ridiculed in his poem "The Way Things Are." He said,

> . . . I'll hound you
> Until you stop insisting things are made
> Of particles like themselves. That's foolishness,
> Sheer lunacy. Surely, a man can laugh
> And not be made of laughing particles. . . .

And in another place he quipped,

> . . . if men
> Are made up of homunculi, farewell

To any sensible theory of atoms.
The little men inside you will be roaring
With mirth till tears run down their cheeks.

Lucretius gave no substantial reasons for ridiculing likepartedness. He rejected the idea out of hand because he thought that the atomic theory—matter composed of small particles unlike anything we can see—did the job better.

Galileo found an argument for rejecting likepartedness, and although he published it a hundred years before Swedenborg we have no evidence that the Swede was aware of its implications. Galileo pointed out that size *does* make a difference. If God were to create elephants ten times as tall as those we know, they would fall in a heap. Such elephants would weigh 1000 times as much (10 x 10 x 10) and would only be 100 times as strong, assuming that they were proportioned after ordinary elephants and that strength was determined by the *area* of a muscle rather than by its volume.

Large and small cannot be alike. As Julian Huxley said, "It pays to be the right size." Objects of vastly different size behave in drastically different ways. Ants, for example, are not hurt by falling from a tree-top, because they are small and light; air resistance on their surface easily slows them down. A large "ant" would be killed, however. A solar system is *nothing like* an atom, despite the fact that both have a central body and smaller objects fly about. And if we imagine that the stars of the Milky Way form a gas—perhaps representing the air inside the lungs of some vast man—we must admit that the gas is nothing like our air on earth, and the vast man cannot be very human.

Of course, all this is not to deny that larger structures are built from smaller ones or even that objects can combine to form similar objects. A "nation" is one such conception: an aggregation of people forming an entity with a personality and a behavioral pattern that can be analyzed and used as the basis for international politics. But this approach is dangerous if taken too seriously, and it tends to miss the point. Just as a nation is more than a mega-person, so a planetary system is more than—and essentially different from—a mega-atom.

Likepartedness is too constrictive an attitude to encompass experience, and it was the realization of this fact that drove men like Galileo and Newton to mathematical formulations which were divorced from everyday experience. Only by turning to mathematics and setting aside the essentially religious search for "meaning" could those scientists bring unfettered imaginations to bear on the problems confronting them.

But Swedenborg was not of the modern ilk; he insisted on like-partedness, and from this conception he developed a description of planetary births. According to Swedenborg, the opening scene of the birth of the planets found the sun in the grip of a vast vortex. He says: "The sun, therefore, is now in possession of its vortex, and the vortex of its sun; both together constituting a system; for one cannot subsist without the other. Hitherto, however, the sun reigns, as it were, alone in a vast court. His empire stretches far and wide, but there are no inhabitants to serve and obey him, to whom he can issue his commands and give laws and enactments."

He then imagines that the outer layers of the sun became compressed and solidified, "throwing into shadow the whole of the world-system, darkening it as if by an extremely opaque cloud." In this state, surrounded by a crust, the sun was "as it were, in a pregnant state, and ready to introduce something new into the vortex."

Swedenborg then lapses into likepartedness, and compares the pregnant sun to an atom of matter, saying:

> The whole of this immense crust, together with the enclosed solar space, is not unlike an elementary particle; for in each elementary particle there is an active space, exteriorly to which flow the finites. Thus, both as to figure and motion, this chaos is, on an immense scale, an effigy of each individual part of an elementary particle. Thus nature is similar to herself in her largest as well as her least productions; and thus she appears in her most stupendous proportions, as well as in her most minute.

He then supposes an acceleration of the rotation of the solar vortex and an outward thrust of the crust. This is the point at which Swedenborg's cosmogony differs from most of the later theories. To postulate an acceleration of the vortex raises more questions than it answers, because it violates the laws governing material vortices; it supposes a new type of behavior.

The crust, thus expelled, then fractures—and the planets form from the fragments.

> The solar crust, being somewhere broken up on the admission of the vortical volume, collapsed upon itself . . . so that it surrounded the sun like a belt or broad circle. This belt, which was formed by the collapse of the incrusting expanse, revolved in a similar manner; removed itself to a greater distance; and by its removal became attenuated till it burst, and formed into larger and smaller globes; that is to say, formed planets and satellites of various dimensions, but of a spherical form.

Depending on their various fates, the fragments of the crust contributed to the planets, the satellites, and the asteroids—or to smaller planets.

Swedenborg concludes by pointing out that variable stars and novae give evidence of dark crusts surrounding stars and breaking loose from time to time.

The essential features of Swedenborg's theory were the formation of a crust about the sun and an acceleration of the solar rotation leading to an expansion and fracture of the crust. This is bad physics—the crust can only accelerate by *contracting,* unless an external force impels it to rotate more rapidly. Swedenborg assumes that the solar vortex will provide this additional impetus, but this should be seen as an arbitrary assumption.

Another difficulty, which has not really been resolved even in modern cosmogonies, is the solidification of the solar matter into the bodies of the planets. How can a gas floating in space condense into a solid? Evidently the planets grew in stages, small condensations accreting more matter, then cooling and coalescing into larger accretions. But there is no possibility of the formation of a solid crust within the atmosphere of the sun; the matter simply cannot cool sufficiently. Smaller bodies could solidify, but the sun is too large to solidify.

It does not appear that Swedenborg's writings on cosmogony had much influence on his successors. Perhaps this was because cosmogonists rarely take each other seriously—and *that* may be the inevitable disappointment of those who try to rationalize origins.

9 MORE SPECULATIONS ON THE NEBULOUS PATCHES

Thomas Wright, an Englishman who raised himself from humble origins to the status of "gentleman" by surveying estates and teaching mathematics and physics to "noble ladies," was born in 1711. As a young man he published several pamphlets on astronomy, architecture, and antiquities, but he is best known for his book *An Original Theory, Or New Hypothesis of the Universe,* which appeared in 1750. It was largely ignored by his contemporaries, perhaps because they did not know what to make of it.

Wright shared one attribute with Isaac Newton: both men were "astrotheologians" who saw the heavens as the revelation of God. But Newton was primarily a scientist, and Wright was primarily a theologian. Newton wished to discover God by studying the universe; Wright wished to discover the universe by studying God.

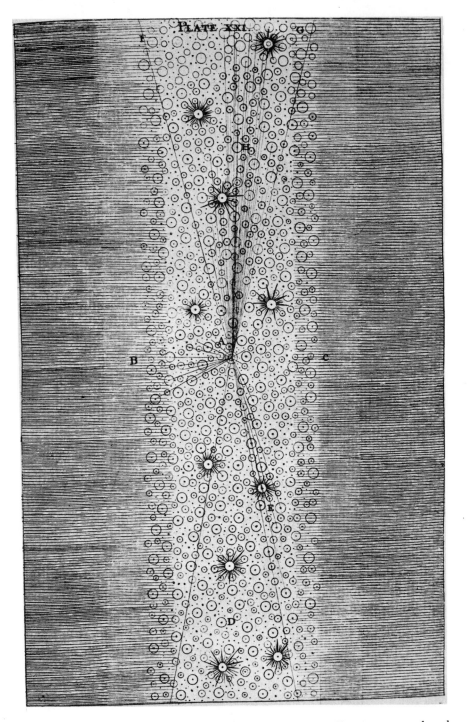

Thomas Wright's model of the Milky Way. This diagram, reproduced from Wright's book *An Original Theory or New Hypothesis of the Universe* (1750), depicts the stars arranged in a slab. Lines of sight from the sun are shown to pass a greater number of stars when they pass at small inclinations to the slab. Compare with the plate on page 80.

At the age of twenty-three, Wright prepared a manuscript putting forth a theory that strongly resembled the medieval motif: shells containing the sun and stars, black emptiness beyond, and a fixed center from which flowed the physical order of the universe. Moral order also flowed from this center. Thus, according to Wright, the structure of the material universe was framed on the structure of the spiritual, or moral, universe. (Swedenborg no doubt would have objected to this, because for him the spiritual universe was framed in the image of man and so it could not also serve as a model for the material universe.)

Wright discussed in his *Original Theory* the philosophical basis for reaching out into the unknown, but he sounds self-conscious, as though he were embarrassed in the face of the sky and feared he would misinterpret the stars unless he kept the image of God before him at all times.

He accepted Galileo's conclusion that the Milky Way is a mass of unresolved starlight. This conclusion, Wright noted, had been held by Democritus, who had believed the Milky Way to be composed of stars "long before astronomy reaped any benefit from the improved sciences of optics; and [who] saw, as we may say, through the eye of reason, full as far into infinity as the most able astronomers in more advantageous times have done since, even assisted with their best glasses." Wright evidently believed that he too could see through the "eye of reason," and his book contains only a cursory discussion of the empirical evidence concerning nebulous stars. The book presents two alternative models of the galaxy, with which "not only the phenomena of the Milky Way may thus be accounted for, but also all the cloudy spots, and irregular distribution of them."

Wright's cosmology was based on three tenets:

1. Our Milky Way is merely one among many in the universe.
2. It appears as a bright band in the sky because it is a flattened layer of stars, at least in our neighborhood.
3. Each Milky Way is concentric to its own supernatural center.

Wright ingeniously constructed two geometrical models, each giving the appearance of a band in the sky and providing a location for the supernatural centers. He did not declare a preference for one over the other.

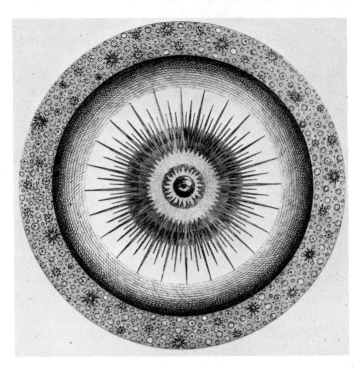

In one of his alternative hypotheses, Wright supposed that the Milky Way, as we see it, is a small portion of a spherical shell. This construction appealed to him, because it provided a natural center for the source of spirit and life, depicted by him as an eye.

In one model, shown in the plate above, the earth is located in a thin shell of stars forming a sphere; at the center is the "eye" of the supernatural. A universe populated with these spheres is like a room full of ping pong balls (plate, page 81), and Wright does not satisfactorily explain how such an array could resemble the observed scattering of nebulous stars.

His other model is more satisfactory, although less elegant. The Milky Way is a ring, or a set of concentric rings like the rings of Saturn; the center is empty to provide room for the supernatural.

Wright did not discuss the possibility of an observational choice between these models, but later writers consistently rejected the spherical model and they transformed Wright's concentric rings into a full circular disk. In this way Wright became known as the first to suggest the disk model of the Milky Way—the model that is now accepted. But Wright's story does not end there.

Wright's universe of galaxies. Each galaxy is a spherical shell, and the shells are packed together to fill space. Distant galaxies produce the nebulous stars according to this theory, which Wright later rejected.

In 1966, a London bookdealer bought eight hundred folios of previously unknown Wright manuscripts and asked the historian Michael Hoskin to assist in arranging them. From the chaos emerged a "wholly unsuspected sequel to the *Original Theory,* entitled *Second or Singular Thoughts Upon the Theory of the Universe."* This manuscript was evidently completed about twenty years after the *Original Theory,* and Wright put forth a totally different view. He rejected all three of his previous tenets and placed the earth within a spherical cavity—the sky was once again solid. The stars were distant volcanoes; the nebulous stars and comets were volcanic eruptions floating above the black ground of the sky. Hoskin published the manuscript with the comment

> Seen from the modern standpoint, these *Second Thoughts* are wholly retrograde: they involve a return to the solid heavens of medieval astronomy, and the abandonment of Wright's famous explanation in the *Original Theory* of the Milky Way as an optical effect due to our immersion in a layer of stars. But seen within the context of Wright's life work, they mark the culmination of his efforts to reconcile the moral and the physical view of the world.

Wright had evidently been dissatisfied with the pluralistic appearance of the *Original Theory,* and he noted that the theory said nothing about new stars and comets. A violent earthquake in 1755 set him to thinking about the interior of the earth, where he imagined fire and liquid rock cavorting and crashing into volcanoes and earthquakes—a scene originally set forth by Edmund Halley. Wright came to conceive of man living within a cavity, and the germ of his second thoughts, in which the Milky Way is a "vast chain of burning mountains forming a flood of fire," is evidently to be found in the book *Astronomical Principles of Religion,* published in 1717 by W. Whiston. This book described stellar structure in the following phrases:

> If there be any such Cavities and Recepticles for living creatures, and the Things necessary for their Sustenance, in the Central Regions of the Sun, of the Planets, or Comets, 'tis certain their state and Circumstances must be very different from those on the Surfaces of the Planets. They must all live in Concave Spheres, which must hinder all Intercourse between them and this visible World: Nor can they have any Philosophical Evidence that there is such an External World at all; which is the case of the rest of this Universe, as to us, if we, with

all the visible Stars, Comets, and Planets, be our selves included in such a Cavity; which is not absolutely impossible to be suppos'd.

While Wright was spawning his poetic vision, the young Immanuel Kant was also at work.

Kant was born in 1724 and lived his entire life in Königsberg, in old Prussia. The son of a harnessmaker, he was the fourth of nine children. The children were raised in the warm atmosphere of a pious and poor family, and although Kant did not acquire the religious faith of his parents, he wrote of their piety:

One may say of pietism what one will; it suffices that the people to whom it was a serious matter were distinguished in a manner deserving of all respect. [My parents] possessed the highest good which man can enjoy—that repose, that cheerfulness, that inner peace which is disturbed by no passions. . . . Even the mere onlooker was involuntarily compelled to respect. I still remember how once disputes arose between the harness-making and saddler trades regarding their privileges, during which my father suffered much. But, nevertheless, this quarrel was treated by my parents, even in family conversation, with such forbearance and love towards their opponents, and with so much trust in Providence, that the memory of it, although I was then a boy, has never left me.

Kant's mother died when he was thirteen, and three years later he entered the University of Königsberg as a student of theology, falling at once under the spell of mathematics and natural science. He became a personal friend of a young professor, Martin Knutzen, who was an ardent proponent of Newtonian physics—and therefore an opponent of the school of Descartes. Kant entered the camp of the Newtonians, and the scientific pamphlets he wrote as a young student demonstrate the excellence of his training. A newspaper account of Wright's *Original Theory* triggered a four-year concentration by Kant on the interpretation of the Milky Way and the nebulous stars, culminating in the publication of his *General History of Nature and Theory of the Heavens* in 1755.

As published in the Königsberg newspapers, the account of Wright's work was fragmentary and misleading—it lacked illustrations and gave the impression that Wright had proposed a disk of stars. Kant accepted the disk model and infused into it the physics that Wright had ignored. Arguing by analogy with Saturn's rings, which he assumed to be swarms of particles rotating about the planet, Kant proposed that the shape of the nebulous stars and of the Milky Way might be explained by their rotation. Focusing on the elliptical nebulosities similar to the Andromeda Nebula (plate, page 55), he argued that such patches of light "can be nothing else than a mass of many fixed stars." His argument was based on the assumption of uniformity in the universe—the opposite of Mairan's earlier assumption that the universe must contain "monsters." Kant wrote:

> As these nebulous stars must undoubtedly be removed at least as far from us as the other fixed stars, it is not only their magnitude [size] which would be so astonishing—seeing that it would necessarily exceed that of the largest stars many thousand times—but it would be strangest of all that, being self-luminous bodies and suns, they should still with this extraordinary magnitude show the dullest and feeblest light. It is far more natural and conceivable to regard them as being not such enormous single stars but systems of many stars.

Kant makes a very important point in his discussion of nebulae—one that had been implicit in Mairan but was certainly absent in Wright—namely, that different types of nebulosity probably require different types of explanations. He admitted that his theory might not apply to objects in which nebulosity and stars appear together (plate, page 64). Kant said, "Here a wide field is open for discovery, for which observation must give the key. *The nebulous stars, properly so called,** and those about which there is still dispute as to whether they should be so designated, must be examined and tested under the guidance of this theory."

It is difficult to construct a down-to-earth parallel to Kant's model of the Milky Way because the gravitational force among the stars is so different from any force we experience among earthly objects. Imagine a flock of birds all tied to each other by elastic strings and constrained to fly parallel to the surface of the earth. If, after some frantic moments of collision, they learn to collaborate in flying circularly about their collective center, they will form a swirl that has some of the attributes of Kant's model for the Milky Way: discrete stars orbiting about a common center in a flattened circular array that will look smooth and

---

* Italics mine, to call attention to a similar phrase in the later writing of Herschel. *Cf.* p. 113

elliptical if seen from a great distance. (A swarm of gnats behaves in much the same way.) Just as there is no need for a fixed point at the center to which all the strings are tied, so there is no need to imagine a fixed center of the spiraling Milky Way. In fact the center might be a figment of our mathematical imaginations, an empty point toward which the accumulated effect of all the stars appears to attract each individual star.

Kant's theory was a master stroke of physical insight. His discussion was incisive; he knew few enough of the observational details so that he was not confused by reality, yet he had all the large facts at his disposal. This is surely among the very finest scientific books of the eighteenth century. Yet, it was published anonymously; worse, the publisher immediately went bankrupt and the book was locked in a warehouse. Inevitably, Kant's ideas were not well known.

In 1761 the astronomer J. D. Lambert, a compatriot of Kant's, published a theory of the Milky Way which was geometrically similar to Kant's but which was less specific and at the same time less complete in its physical details. In 1763, Kant published a summary of his theory in a preface to his book *Only Possible Proof of the Existence of God.* Lambert, discovering that he had been preceded, took the news with abundant grace, and he explained the circumstances of his own publication in a letter to Kant, dated 1765:

> I can tell you with confidence, dear sir, that your ideas about the origin of the world, which you mention in the preface to *Only Possible Proof* were not known to me before. What I said on page 149 of *Cosmological Letters* [1761] dates from 1749. Right after supper I went to my room, contrary to my habit then, and from my window I looked at the starry sky, especially the Milky Way. I wrote down on a quarto sheet the idea that occurred to me then that the Milky Way could be viewed as an ecliptic of the fixed stars, and it was this note I had before me when I wrote the *Letters* in 1760.

Lambert goes on to say that he had not heard of Wright's ideas until 1761, after the completion of his own book. It seems, then, that the three men—Wright, Kant, Lambert—developed their ideas independently of each other, although Kant was drawn to consider the Milky Way and the solar system by news of Wright's book.

In any case, by the time he heard from Lambert, Kant was no

longer interested in the problem; Lambert tried to draw him into correspondence on the subject, but Kant's mind had already turned inward to
a reflection on human knowledge and the process of reasoning, so he let
his cosmology rest. Because the original book had been published
anonymously, even the fame growing out of his philosophical works did
not draw attention to his theory of the universe. Later texts, for example, by Johann Erxleben, mentioned Wright and Lambert, but not
Kant.

In 1791, Kant was aroused once again to defend his priority.
William Herschel, who had already published several papers on the
structure of the Milky Way and nebulosities, announced that he had
been able to discern the rotation of the inner ring of Saturn. The supposed rotation of this ring had been one of the key analogies in Kant's
discussion of the Milky Way, and he at once sent a letter to the editor of
the German publication *Astronomical Yearbook*. He wrote the following:

> If we are to understand what I recently read in a political newspaper,
> namely that Mr. Herschel has discovered a rotation of the innermost
> edge of Saturn's ring in 10^h 22^m 15^s, this would confirm what I as-
> sumed 35 years ago in my *General History of Nature and Theory of
> the Heavens,* namely that this part of the ring is floating freely in
> circular motion according to the central force law (for the inner edge
> I computed on page 87 an orbital time of 10 hours). Also, Mr. Her-
> schel's concept [*Vorstellungsart*] of the nebulous stars to be systems as
> such and also in a collective system, agrees happily with what I pro-
> posed at that time, and it must be a slip of memory of the late Erxleben
> that in his physics, he ascribed this thought to the late Lambert, who
> is said to have had it first, since Lambert's *Cosmological Letters* ap-
> peared 6 years later than my work, and also, despite all my searches in
> these letters, I have not found that concept.

The following year Kant permitted selections of his *General
History* to be reprinted, along with a German translation of three
treatises by Herschel; in 1798 the entire work became available.

Thus, 180 years after Galileo's first use of the telescope, the nebulous stars were still the subject of speculation. Of the men who took the
problem seriously, none had adequate telescopes; of the men who could
have constructed adequate telescopes, all were occupied with other
problems; none devoted themselves to these peculiar objects.

10 WILLIAM HERSCHEL TURNS TO ASTRONOMY

William Herschel, a German oboist who deserted the army and emigrated to England, opened a new era of astronomy. The magnitude of his accomplishments and the labors they required are awe-inspiring; he has been called the greatest observational astronomer, and I can think of few men in competition for the title. For thirty years Herschel swept the sky with telescopes whose grandeur of conception and construction were unique, and remained unique for two decades after his death in 1822. Yet Herschel had not known an astronomer nor built a telescope until he was thirty-five years old.

Herschel must stand with Galileo and Newton as a creator of modern astronomy; but the three men stand on the far points of a triangle if we compare their lives and attempt to guess at the sources of their creative urges. Reading Manuel's biography of Newton, I picture a painfully lonesome child who found solace in mechanical creations and grew into an isolated adult who discovered he could create a world of mathematical certainty and thus expose the wonder of his Father in heaven. To the extent that Newton was indeed motivated by this discovery of personal identity in a mathematical exposition of God's work, he is similar to another great man, René Descartes, who found the basis for his sense of identity in the mere fact of his rational self-awareness, and who then constructed a philosophy in the spirit of a mathematical exercise, starting on the bedrock of his personal intuitions and stepping carefully forward along a precisely determined path.

By contrast, Galileo found his reality in much the way of the existentialists three centuries later—not by sublimation of his own need for certainty in the certainty of a physical universe, but by assimilating himself into the universe and finding identification as a part of the world. Galileo was one of the earliest of the modern protean men; he had no fixed idea of himself apart from the world. He explored the world to understand himself, not—as seems to have been the case with Newton—to create an intellectual cathedral in which to enthrone the divinity (a cathedral from which even the human architect seemed excluded). As Lionel Trilling implies in his book *The Experience of Literature,* the modern playwright Bertolt Brecht has succeeded in showing the nature of Galileo by creating an opposite to what Galileo must really have been. In the play *Galileo,* we see a nothing; a transparent, taciturn man who lives in a misty world of dreams—as far removed from Galileo's world as one can imagine.

The fact that Galileo dealt in experiments and Newton dealt in mathematical synthesis may have been a consequence of their times: Newton arrived when the advance of astronomy needed synthesis, not experiments or observations. But to say that the bents of these two men differed because their times differed would be to underestimate the powerful imprint of a man on his time: great scientists make their times, they do not merely respond to them. To put it in the words of the French philosopher Bergson, the real professionals of an era define the questions and formulate the answers, they are not satisfied to answer the questions asked by men before them; that is the task of amateurs. I have the impression that Galileo sought the world with the same courageous love he showered on his friends, while Newton held the world in his slender fingers and saw it—saw through it—as no other man had done; his was an achievement of determined intellect.

Herschel combined some elements of both men, and the late hour of his flowering is a fascinating story in itself, a story to which the astronomical career of his only son adds a suggestive footnote.

William was born in 1738 to Isaac Herschel, who had turned from agriculture, his father's profession, to music. Isaac had given up positions in two regimental bands, one in Berlin which he said was "very bad and slavish" and one in Brunswick which was "too Prussian," before settling down in Hanover as an oboist in the Foot-Guards. Virtually all of Isaac's children developed musical talent under his tutelage and that of the tutors he hired. William joined the regimental band at the age of fourteen and his older brother Jacob joined even earlier.

The Herschel household must have been a very stimulating one for the young boy. His younger sister Caroline later wrote that her

> brother William and his Father were often arguing with such warmth that my Mother's interference became necessary, when the names of Leibnitz, Newton and Euler sounded rather too loud for the repose of her little ones, who ought to be in school by seven in the morning. . . . My Father was a great admirer of astronomy and had some knowledge of that science; for I remember his taking me on a clear frosty night into the street, to make me acquainted with several of the beautiful constellations, after we had been gazing at a comet which was then visible. And I well remember with what delight he used to assist my brother William in his philosophical studies, among which was a neatly turned globe, upon which the equator and the ecliptic were engraved by [William].

What with "fiddling," playing the hautboy, studying languages, mathematics, and philosophy, William received a wide if not altogether coherent education. An early kindness toward Caroline, twelve years his junior, is related in her memoires: "My Mother being very busy in preparing dinner had suffered me to go to a parade to meet my Father, but I missed him, and continued my search till I was spent with cold and fatigue; and on coming home I found them all at table; nobody greeting me but my brother William, who came running and crouched down to me, which made me forget all my grievances."

Caroline contracted smallpox at the age of four, and she later wrote, "Although recovered, I did not escape being totally disfigured and suffering some injury to my left eye." Her love for William, which later became an important element in their lives, may have been nourished by her sense of disfigurement. Caroline never so much as hinted at a criticism of William, although she was not above censuring her other brothers.

Shortly after the great Lisbon earthquake in 1755, the three regimental Herschels accompanied their group to England, where they stayed six months, making acquaintances which proved valuable later, and learning the English language. With some difficulty, and to avoid borrowing from his father, William managed to save enough money to buy Locke's essay *On Human Understanding,* with which he entertained himself. Jacob obtained release from the regiment and returned to Hanover ahead of the others, who arrived in the autumn of 1756. The next spring, according to William's *Biographical Memoire,* his father and he

> went with the regiment into a campaign which proved very harassing by many forced marches and bad accommodations. We were obliged, after a fatiguing day, to erect our tents in a ploughed field, the furrows of which were full of water.
>
> July 26. About the time of the battle of Astenbeck we were so near the field of action as to be within reach of gunshot; when this happened my Father advised me to look to my own safety; accordingly I left the regiment and took the road to Hanover; but when arrived there I found that, having no passport, I was in danger of being pressed for a soldier; it was therefore thought proper for me to return to the army.
>
> When I had rejoined the regiment I found that nobody had time to look after the musicians; they did not seem to be wanted. The weather was uncommonly hot and the continual marches were very harassing. At last my Father's opinion was that as, on account of my youth, I had not been sworn in when I was admitted to the Guards I might leave the military service; indeed he had no doubt but that he could obtain my dismission; and this he afterwards procured.

William met Jacob in Hamburg—Jacob had been under cover to avoid being conscripted—and the two embarked for England, virtually penniless but assured of freedom. They landed in November of 1757 and went straight to London, where William obtained work copying music. The two brothers had lived in London for two years when Jacob was offered a place in the orchestra in Hanover, and he eagerly accepted. Left alone, William pursued music with a vengeance; he hoped to distinguish himself as a composer, and he delved deeply into music theory while he played the violin and organ in local concerts. In 1761 he wrote that in the three years spent in England he had not met a single person whom he felt worthy of friendship; a month later he wrote Jacob a bantering letter in French describing a young girl who delighted him with her frank and open friendliness, unspoiled by delicacy or reserve. But he was still thirty years away from marriage.

The philosophy that Herschel had acquired by this time (he was in his early twenties) is suggested by this passage in a letter to Jacob:

> There are two kinds of happiness or contentment for which we mortals are adapted; the first we experience in *thinking* and the other in *feeling*. The first is the purest and most unmixed. Let a man once know what sort of a being he is; how great the being which brought him into existence, how utterly transitory is everything in the material world, and let him realize this without passion in a quiet philosophical temper, and I maintain that then he is happy; as happy indeed as it is possible for him to be. . . .
>
> Where the attainment of our wish is accompanied by a certain melancholy, we must remind ourselves that this present life cannot be perfect; or more truly we may say that this sadness is in itself a sign that we have reached the furthest stage of happiness which our sensibilities are capable of giving us. This leads us with a kind of yearning towards our Creator. It is He Himself whom we desire and this mixed feeling of sadness and of joy comes from the thought of the perfection of the Creator and of the dependence and imperfection of the creature.

In 1762, William was appointed director of public concerts in Leeds and his life became a bit more settled, but simultaneously he showed the first signs of restlessness with music. To his brother he wrote, "It is a pity that music is not a hundred times more difficult as a science. . . . My love for activity makes it absolutely necessary that I should be busy, for I grow sick by idleness; it kills me almost to do nothing."

Isaac Herschel died in 1767, after three years as a partial invalid.

Caroline was seventeen and was miserable in the company of her mother and older brother—her younger brothers were her only solace. Even her father, well-meaning and pious, had not added to her later comfort, as we may gather from this entry in her autobiography:

> I saw that all my exertions would not save me from becoming a burden to my brothers; and I had by this time imbibed too much pride for submitting to take a place as a Ladiesmaid, and for a Governess I was not qualified for want of knowledge in languages. And I never forgot the caution my dear Father gave me; against all thoughts of marrying, saying as I was neither hansom nor rich, it was not likely that anyone would make me an offer, till perhaps, when far advanced in life, some old man might take me for my good qualities.

Small wonder that William was able to persuade Caroline to join him in England. Dietrich, the youngest brother, had already moved in temporarily with William; and Alexander, the penultimate one, had taken to "associating with young men who led him into all manner of expensive pleasures." Soon Alexander emigrated to England at William's invitation, and in 1772 their mother's objections were overcome and Caroline was permitted to join William in Bath, where he had a half-house in which he installed Alexander and Caroline (Dietrich by this time having returned to the Continent).

Caroline was at first lonesome among the English women, most of whom she considered to be "idiots," but she soon became an accomplished singer, thus joining the musical round with William.

On May 10, 1773, William bought a copy of Ferguson's *Astronomy,* the leading text of the time. He said, "When I read of the many charming discoveries that had been made by means of the telescope, I was so delighted with the subject that I wished to see the heavens and Planets with my own eyes thro' one of those instruments." Caroline told how he "used to retire to bed with a bason of milk or glass of water, and Smith's *Harmonics* and *Optics,* Ferguson's *Astronomy,* etc., and so went to sleep buried under his favorite authors; and his first thoughts on rising were how to obtain instruments for viewing these objects himself." The texts, it would seem, were more tantalizing than informative in matters of the fixed stars. Smith's *Optics* declared that "the more of them [could be seen] as the aperture is more enlarged to take in more light," and Herschel determined to make immense telescopes to explore the sky. Ferguson had written, "There is a remarkable tract round the Heavens called the Milky Way from its peculiar whiteness, which was formerly thought to be owing to a vast number of stars therein; but the telescope shows it to be quite otherwise; and therefore its whiteness must be owing to some other cause."

Within months of William's purchase of these books (and as soon as his music students left him for the summer), the house was converted into a workshop, and Caroline and Alexander were pressed into service. The first efforts with lenses of modest dimensions were rather disappointing, were extremely awkward to use, but Herschel bought some used mirror-making equipment from a neighbor, and during the autumn the neophyte received lessons in the grinding and polishing of mirrors.

With each successful observation of a planet, with each telescope completed and turned on the sky, Herschel's enthusiasm mounted. Music began fading into the background. For Caroline, the alteration of their way of life brought new meaning to her own, and she was not slow to realize the possibilities. Of the summer of 1774 she wrote:

> During this summer I lost the only female acquaintances (not friends) I ever had the opportunity of being very intimate with, by the Bulmans' Family returning again to Leeds. For my time was so much taken up with copying Music and practising, besides attendance on my Brother when polishing, that by way of keeping him alife I was even obliged to feed him by putting the Vitals by bits into his mouth;—this was once the case when at the finishing of a 7 feet* mirror he had not left his hands from it for 16 hours altogether. And in general he was never unemployed at meals, but always at the same time contriving or making drawings of whatever came into his mind. And generally I was obliged to read to him when at some work which required no thinking; and sometimes lending a hand, I became in time as useful a member of the workshop as a boy might be to his master in the first year of his apprenticeship.

From 1773 to 1782, the Herschels went through a busy period of transition from professional musicians to professional astronomers. At the start of this interval, William had only his spare moments for telescope making and observing; at the finish, when he was forty-three, he

---

* When telescopic dimensions are quoted in feet, they refer to the focal length of the mirror; when they are quoted in inches, they refer to the diameter of the mirror.

had been appointed astronomer to the court of George III and had taken up residence in Windsor; music was relegated to his spare moments. Herschel would have been the first to acknowledge that the indulgence of his brother Alexander and his sister Caroline made the transition possible, and they in turn must have realized that their collaboration with William, who remained simple and generous through the peak of his fame, was the inspiration of their own lives. Caroline became Herschel's chief observing assistant, recording observations at the telescope through damp English nights and transcribing them the next day into catalogues. Until 1788, when Herschel married, she managed the household as well. Needless to say, his marriage was a bitter blow to Caroline, although the only real evidence we have of her thoughts is the fact that she tore from her journal the pages covering the decade after the event. Mrs. Herschel's warmth, their mutual love of William, and the delightful friendship of John, the only child of the marriage, finally brought the two women together into graceful old age.

Caroline became known among the astronomers of England and Europe as an astronomer in her own right—probably the first woman thus allowed into the ranks. One week, while William was on the Continent and Caroline had some free time, she handled the telescope by herself and discovered a comet—the first of her several. Even today the discoverer of a comet is applauded the way a person who catches a prize fish might be; in those days it was yet more commendable—I suppose because no one knew quite what significance to attach to comets, and because comets brought unpredictable life to the sky. Just as publishing a poem is said to make a man or woman a poet today, so discovering a comet made a man or woman an astronomer then.

The events leading to Herschel's coming under the wing of the King commenced in December 1779, while the astronomer was looking at the moon from the street before his house. He wrote that "a gentleman coming by . . . stopped to look at the instrument. When I took my eye off the telescope he very politely asked if he might be permitted to look in, and this being immediately conceded, he expressed great satisfaction at the view." This man, William Watson, Jr., later provided Herschel's introduction to the local Literary Society, the Royal Society of London, and finally to King George III.

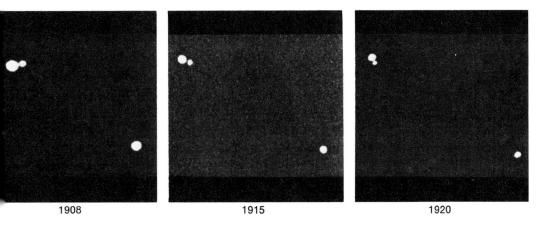

1908   1915   1920

The binary star Krueger 60, showing the orbital motion in a period of twelve years. The third star in these photographs is not known to be a member of the system. Motions such as these were discovered by William Herschel at the end of the eighteenth century, and they provided convincing evidence that gravitation operates among the distant stars. (*Yerkes Observatory, University of Chicago*)

The telescope through which the gentleman was moonstruck was the seven-foot reflector with a four-inch mirror, the same with which Herschel completed his first survey of the sky, touching on all of the bright stars. The first systematic project that attracted Herschel was to study Mira, the only star whose brightness was then known to vary. He published one paper on that star and then looked for more fertile fields of research. In a preliminary survey of the sky he discovered a number of stars that were closely paired in the sky—double stars, such as the one illustrated above. In most such pairs, one star was several times brighter than the other, and Herschel realized that if the fainter star were at a greater distance he might be able to detect the motion of the sun as a relative motion of the two stars, in much the way that the distant trees appear to move back and forth relative to nearer ones when we nod our head. Also, if the fainter star were much farther than the brighter, a simple calculation based on a diagram such as that on page 96 might permit measurement of the nearer star's distance. No such measurement had yet been achieved despite numerous attempts. In fact no star had yet had its brightness determined—astronomers could still do no better than assume that the stars were of roughly the same brightness as the sun. So Herschel undertook a systematic search

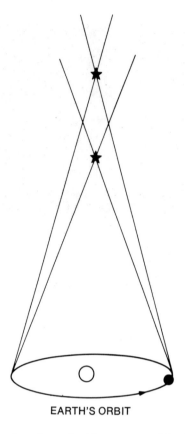

EARTH'S ORBIT

As the earth moves in its orbit about the sun, the more distant stars will provide a background against which to measure the motion (called "parallax") of the nearby stars. Herschel hoped to measure stellar distances in this way, but instead he found that the stars in close pairs revolved about each other.

for close double stars, and he began the preparation of a catalogue and list of measurements.

In 1781 he discovered Uranus and became unique in the history of man—the first astronomer to discover a primary planet. The discovery was not accidental. As he himself said:

> It has generally been supposed that it was a lucky accident that brought this star [a misnomer which was common at the time] to my view; this is an evident mistake. In the regular manner I examined every star of the heavens, not only of that magnitude but many far inferior, it was that night *its turn* to be discovered. I had gradually perused the great Volume of the *Author of Nature* and was now come to the page which contained a seventh planet. Had business prevented me that evening, I must have found it the next, and the goodness of my telescope was such that I perceived its visible planetary disc as soon as I looked at it.

So far so good; he later proved many times over that his telescopes were indeed superior to all others. Messier was astonished at Herschel's sharpness. He wrote, "Nothing could be more difficult than to recognize it, and I cannot conceive how you were able to return several times to this star—or comet—as it was absolutely necessary to observe it several days in succession to perceive that it had motion, since it has none of the usual characteristics of a comet."

But Herschel, as did most astronomers who read his first report, naturally assumed he had discovered a comet. Comets usually appear to move more swiftly than planets, and this fact may account for his spurious detection of motion during the first few hours of observation—

The planet Uranus. This photograph is a 3½-minute exposure with the 82-inch telescope of the McDonald Observatory, Texas. It was made in 1948, and it shows the disk of the planet crossed by lines of light produced within the telescope and surrounded by a ring of light, produced within the photographic plate. The most recently discovered satellite of Uranus, known as the "Vth," may be seen within the ring. Its mean distance from the planet is 81,000 miles and its orbital period is thirty-four hours. (*Yerkes Observatory, University of Chicago*)

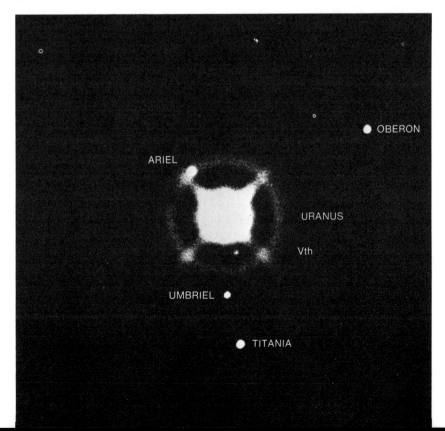

motion that proved to be difficult to live down during the early days of his career. He discovered the "comet" on March 13, 1781, but before it had moved far enough to permit an accurate orbit to be measured, the sun intervened and the object was not seen again until the following August. Preliminary calculations showed the object to have a nearly circular orbit and a period of about twelve years.

Comet or not, it was no ordinary object, and for its discovery Herschel was awarded the annual gold medal of the Royal Society of London and on December 7 was elected to membership in that brilliant constellation.

But as frequently occurs to amateurs who find themselves thrust among professionals, Herschel proved to be a somewhat erratic star. He was not gauche—far from it—but he was naïve in the ways of professional scientists: a single error, or a suspicious conclusion, or an outlandish claim is enough to poison the air about a scientist for years to come. A paper on the motion of the sun, which appears to have been immediately responsible for Herschel's election—or so we may judge from the fact that his election seems to have been delayed until the paper had been read—contained the casual yet fantastic claim that he had successfully employed a magnification of 5400 with his telescope.

Herschel's friend Dr. W. Watson, Jr., wrote to him the day of his election, enclosing a copy of the new catalogue of nebulous stars published by Messier and Méchain, and the following week Watson revealed to Herschel the mixed reactions of the members of the Royal Society:

> Your paper was read at the last meeting; it gained great applause from the generality of those two who were present. Yet I have reason to think that the astronomers were put to a stand for [what] to think of it. Dr. Maskelyne [the Astronomer Royal] is, from what I could find, ready to admit its merit, tho' he said, he did not think the method of computing the interval between two stars by guessing how many diameters of either star it would subtend, a good one: tho' he was ready to own he might change his opinion when he came to read over the paper.* He said likewise some of postula were very disputable,

---

* The method is *not* a good one, as the apparent diameters of stars are properties of the telescope and the earth's atmosphere; they vary from hour to hour and they are quite unreliable as measuring units.

especially that by which the stars of equal size [apparent brightness] were supposed to be at equal distances from us. Mr. Russel, who has not like the rest had an opportunity of knowing more of your merit, told me frankly he thought the paper was a flighty one, so that your prognosis that some would think you fit for Bedlam when you talked of a power of 5400 has been verified. I dined with him yesterday in company with Aubert, Nairne, and others. I found it was the high power you use which prevented him from giving you your due merit. He said, however, that he would suspend his judgment. I told him that was all I required, and that, to use an expression I have heard made use of by Dr. Priestley [a prominent scientist], you desired only a clean stage and no quarter . . . and so, my good friend, you have arrived at such perfection in your Instrument, or at least have dared to apply it to such uses and so have overleaped the timid bounds which restrain modern astronomers, that they stand aghast and are more inclined to disbelieve than to admit such unusual excellence.

Watson's friendly castigation of the modern astronomers is only partly justifiable: Herschel's telescopes were far superior to all others, but under no circumstances is it possible that magnifications greater than, say, 500 could have revealed more detail than lower powers. The nature of light puts an inexorable limit on the amount of detail visible in a telescope, although it is possible Herschel may have been more comfortable using the high powers. In the last analysis, the optimum power is a matter of taste; Herschel's taste clearly differed from that of his contemporaries. Scientifically, the question is relatively unimportant, but the implication—that Herschel might be a crank—was not easily dismissed. As Priestley said, a "clean stage" was what Herschel required, and it was given him by his fellow astronomers because they soon had evidence that Herschel's observations were extraordinary, if his methods and some of his conclusions were a bit difficult to swallow.

As an epitome of Herschel's approach to cosmology, these high magnifications are perhaps more significant. Herschel reached further than his contemporaries; he was determined to discern the construction of the heavens, even though this meant stretching and using postulates that could not be rigorously true. He entered astronomy in middle age, without the conservative training of most professionals, and he was anxious to get on with the work. Probably no other man in history has spent as many hours with a telescope as Herschel. There was no photography then; there were no libraries of celestial atlases depicting

Herschel's 20-foot reflecting telescope, at which he spent years of English nights. The observer rode at the top and was assisted by a person on the ground, who helped to adjust the telescope and to record the observations. The entire structure could be rotated on a track. (*Yerkes Observatory, University of Chicago*)

the faint stars and nebulosites. But there were scattered drawings, and there was the catalogue of Messier and Méchain.

Within eight months of his election to the Royal Society, Herschel was awarded an annual stipend of £200 by King George III. (Herschel then proposed to name his new planet after the King, but tradition prevailed and a name was chosen from mythology.) His duty was to be on hand with a telescope when the Royalty wished to see the sky. One such evening, described in a letter from Herschel to his sister shortly before the royal offer, was probably typical:

This evening as the King and Queen are gone to Kew, the Princesses were desirous of seeing my Telescope, but wanted to know if it was possible to see without going out on the grass; and were much pleased when they heard that my telescope could be carried into any place they liked best to have it. About 8 o'clock it was moved into the Queen's apartments and we waited some time in the hopes of seeing Jupiter or Saturn. Meanwhile I showed the Princesses and several other ladies that were present, the Speculum [mirror], the Micrometers [for measuring angles between stars], the movements of the Telescope, and other things that seemed to excite their curiosity. When the evening appeared to be totally unpromising, I proposed an artificial Saturn as an object since we could not have the real one. I had beforehand prepared this little piece, as I guessed by the appearance of the weather in the afternoon we should have no stars to look at. This being accepted with great pleasure, I had the lamps lighted up which illuminated the picture of a Saturn [cut out in pasteboard] at the bottom of the garden wall. . . .

Tomorrow they hope to have better luck and nothing will give me greater happiness than to be able to show them some of those beautiful objects with which the Heavens are so gloriously ornamented.

Herschel's mammoth 40-foot telescope. This instrument was the culmination of Herschel's telescope building, although most of his important work was done with the 20-footer. Observing with such a device was dangerous and it produced ague. (*Yerkes Observatory, University of Chicago*)

Upon hearing of the King's stipend, Watson wrote to Herschel, "You are now upon the eve of entering into a new course of life, to take upon you a new character, and to be in a situation which will at the same time command respect and what is still more desirable, enable you wholly to give up yourself to an employment which is attended with the highest gratification."

And so indeed he was.

11 HERSCHEL EXAMINES
THE NEBULAE

The catalogue of Messier and Méchain, sent to Herschel on the day of his election to the Royal Society, had the effect of candy set before an infant. He reached for every object on the list, and he was well equipped because he had completed a twelve-inch telescope that autumn and had started toward a larger one. The descriptions of that event leave no room for doubt that astronomy was a hazardous occupation for those who insisted on building their own telescopes. After the metal for a thirty-inch mirror had been poured into the mold, Herschel reports, "We perceived that some small quantity began to drop through the bottom of the furnace into the fire. The crack soon increased and the metal came . . . upon the pavement [and] the flags began to crack and some of them to blow up, so that we found it necessary to keep a proper distance and suffer the metal to take its own course." Caroline wrote that the stone fragments flew "in all directions as high as the ceiling," and that her "Brother fell exhausted by heat and exertion on a heap of brickbatts."

Herschel reacted to the flying stone as a ranch hand might after his first encounter with a wild horse: he immediately built another twenty-foot telescope with an aperture of nineteen inches. This became the workhorse, but he required still more light. The plans for a leviathan developed in his mind, a forty-foot telescope with a mirror of forty-eight inches aperture, and he built it with £4000 of the King's money and a crew of workmen that he personally supervised—but the enterprise was a failure by most reasonable standards. The structure carrying the forty-foot iron tube is shown in the plate on page 101; it was so immense that two workmen were required to assist Herschel when he set on a star, and the process had some of the elements of shaving with a guillotine. Although no one was actually killed, several were hurt—and the stars hardly noticed this monster staring at them.

But try to imagine Herschel in the shadow of Windsor Castle with an "ultimate object" to discern the construction of the heavens. His choice of a starting point was a natural one: his equipment was built for light-gathering, not for precise measurement of interstellar angles, because he wanted to find the grand pattern; so he explored each of the strange objects enumerated by the Frenchmen. He said, "I saw, with great pleasure, that most of the nebulae, which I had an opportunity of examining in proper situations, yielded to the force of my light and power, and were resolved into stars." Like Galileo a century and a half before him, Herschel knew he had a unique instrument, and his enthusiasm swept him to the conclusion that all of the Milky Way and its brighter knots could be resolved into component stars.

Herschel had expected to find several nebulae which had remained undiscovered—he found two thousand within seven years. While his sister sat with him, he dictated brief descriptions of each nebula that his stationary telescope swept from the sky. He determined positions by noting the elevation of the telescope and the time at which the object passed the field of view. The method was far too crude to detect motions of single stars, but Herschel realized that close pairs might reveal tiny displacements within his lifetime. In this he was correct, but the motions they revealed were not at all what he had expected. He intended to use these pairs to measure stellar distances, but he found that many of the stars in pairs were rotating about each other—they were physically bound together; the binding force was gravitation, as the orbits themselves revealed.

This was a grand achievement—to extend Newton's laws to the realm of the stars—but Herschel's gratification must have been mixed with chagrin; one might even suspect, horror. Suppose you were faced with the imperative task of determining the distances of the stars. On what assumption would you build your technique? Would you not accept the idea that all stars might be treated as having roughly the same brightness—that faint stars are distant while bright stars are close? That is precisely what Herschel did; he did it gladly in his early astronomical nights, and did it desperately during his later nights. He assumed that he could outline the Milky Way by counting the number of faint stars in each direction; that number was taken as a direct indication of the extent of the star-filled region, and therefore of the distance to the edge of our own system. The diagram he constructed on this basis is shown in the plate on page 106. It is reminiscent of the models of Kant and Wright, but Herschel hardly could have known of Kant's book, and Wright's would have seemed inconclusive and even self-contradictory to Herschel, so he probably was very little influenced by it. After his early soundings of the sky he diagrammed our stellar system as an enormous grindstone filled with stars.

Accepting the idea that the Milky Way is composed of starlight and that faintness means distance, it requires no feat of the intellect—only a clear night—to imagine that we live near the center of a flattened system of scattered stars. But Herschel's telescope revealed thousands of faint nebulae, and if they were composed of stars they must be very distant fragments of our own system or indeed separate systems external to our own. Many of them were found in the darker regions of the sky away from the Milky Way—in fact Herschel discovered that these nebulae were more dense away from the Milky Way than in it. He concluded that they were outsiders. Thus he came to the conviction that we float in a separate system of stars—in our own galaxy, which we have called the Milky Way. Beyond the limits of our own system he saw uncountable others afloat.

But the equivalence of faintness and distance, one of the concepts by which he was led to this conception, was disputed by his own discovery of double stars, in which one was obviously much fainter than the other although both members were at the same distance from the earth. This discovery would have sufficed to disprove that stars were of equal intrinsic brightness if Herschel had wished it to; but he needed a tool such as the equivalence of faintness and distance, so he rationalized his way out of the corner by pointing out that factors of two or three in the brightnesses of stars were not so terribly important. Since his time other evidence has shown that stars vary by a factor of a million or more from one another.

Herschel's first observation of the objects listed by Messier and Méchain revealed the splendid variety of the nebulae and clusters of stars—as depicted in the plates on pages 107–9 with photographs

Our galaxy according to Herschel. Herschel originally thought that by counting the stars visible in each direction he could determine the outline of the Milky Way. This diagram was the result, but he appears to have abandoned this conception in his later work. The sun is indicated by the large star near the center; the large bifurcation corresponds to the apparent splitting of the Milky Way. (See plate, page 217.) (*Reproduced from* The Collected Works of Sir William Herschel, *published by the Royal Society and the Royal Astronomical Society, London, 1912—Yerkes Observatory, University of Chicago*)

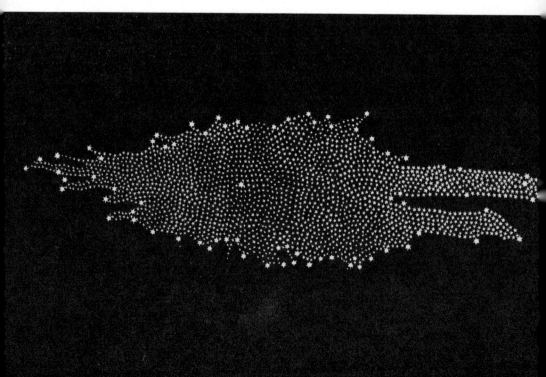

The Great Nebula in Orion. This nebulosity, the brightest in the sky, forms the middle star of Orion's sword. William Herschel described it as "an unformed firey mist, the chaotic material of future suns." (*Yerkes Observatory, University of Chicago*)

Cluster of nebulae in the constellation Hercules. A rich field of the bright dashes and spindles discovered by Herschel. This photograph, taken with the 200-inch Hale telescope, shows nearly as many nebulae as stars.   (*Hale Observatories*)

and, on page 110, taken from his publication. In his own words, "I have seen double and treble nebulae, variously arranged; large ones with small, seeming attendants; narrow but much extended, lucid nebulae or bright dashes; some of the shape of a fan, resembling an electric brush, issuing from a lucid point, others of cometic shape, with a seeming nucleus or in the center."

As time wore on, the astronomical arguments against relating faintness to distance became more persuasive and Herschel came to

---

Stefan's Quintet (N.G.C. 7317–20). The nebulae in this compact group appear to have perturbed each other. They are now known to be galaxies of stars, and they were presumably formed together. (*Lick Observatory, University of California*)

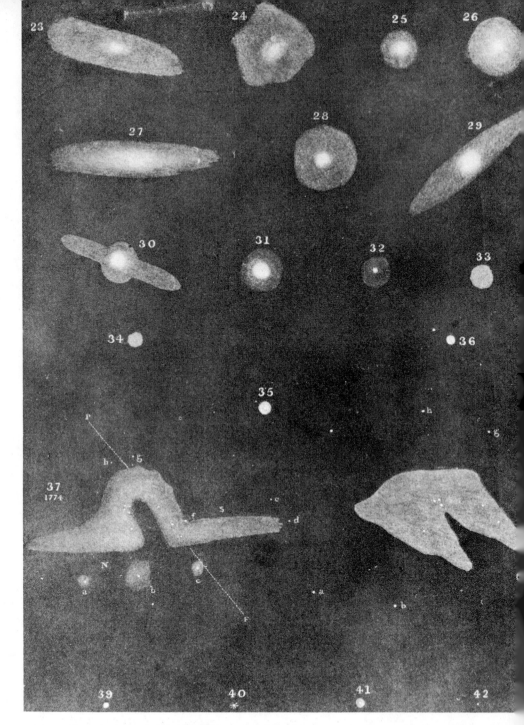

Drawings by Herschel. The two large objects in the lower portion of this plate are separate renditions of the central portions of the Great Nebula in Orion. The difference was illusory, and it illustrates the uncertainties of astronomical observation before the invention of photography. (*Reproduced from* The Collected Works of Sir William Herschel, *published by the Royal Society and the Royal Astronomical Society, London, 1912*)

realize the difficulties of measuring brightness, so he turned from the spatial description of the Milky Way to a chronological description. He sought a scheme on which to hang the forms he had discerned; he entered a new domain of astronomical inquiry: What evidence does the sky hold for the evolution of the universe? In 1789 he provided an analogy which still illuminates astronomical techniques:

> This method of viewing the heavens seems to throw them into a new kind of light. They are now seen to resemble a luxuriant garden, which contains the greatest variety of productions, in different flourishing beds; and one advantage we may at least reap from it is that we can, as it were, extend the range of our experience to an immense duration. For, to continue the simile I have borrowed from the vegetable kingdom, is it not almost the same thing, whether we live successively to witness the germination, blooming, foliage, fecundity, fading, withering, and conception of a plant, or whether a vast number of specimens, selected from every stage through which the plant passes in the course of its existence be brought at once to our view?

Although his theoretical ideas were based on Newton's *Principia,* Herschel seems to have retained the ancient view that motion rather than acceleration is the consequence of force, as though the stars moved in a sticky milieu and would not budge unless drawn by gravity. In an early synthesis of cosmology, Herschel started from a uniform, static universe of scattered stars—Newton, in his letter to Bentley quoted in Chapter 6, took a similar view, but he supposed the initial state to be gaseous rather than stellar; and in this Newton is closer to present cosmology than his successor. Herschel thought that globular clusters of stars (plate, page 62) might be condensations of lesser stars surrounding a single large star; irregular clusters might have formed about pairs of triplets or large stars; such groups might be compounded into larger, more complex agglomerations. He concluded these speculations with the comment that eventual destruction of such groups would be compensated elsewhere in the universe by new formations. Thus he asks us to imagine a maintenance of clusters which we might compare to the minting of new money to replace the old.

In one of his early papers (1785), Herschel set aside a class of nebulae which were ultimately to have a profound bearing on his cosmology—they were in fact to throw him into confusion. He called

Dark region in the Milky Way. Such regions of the Milky Way were called "holes," and were originally thought to be gaps in the star clouds. They are now attributed to dark clouds of dust. (*Yerkes Observatory, University of Chicago*)

them "planetary" because they vaguely resembled the disk of a planet and he admitted that he was "in doubt where to class them." They did not fit his scheme, and this paper came to no conclusion.

Herschel suggested that the peculiar objects, and by implication other small nebulae, might serve a practical use in astronomy. If they are outside our own system they "may then be expected to keep their situation better than any one of the stars belonging to our system, on account of their being probably at a very great distance. Now to have a fixed point somewhere in the heavens, to which the motions of the rest may be referred, is certainly of consequence in astronomy." This, as far as I am aware, is the first published suggestion that distant nebulae be used as reference points, and two decades ago they were put to just that use for the first time. Laplace, the French mathematician whose *Système du Monde* we will see to be closely related to Herschel's work, repeated the suggestion without mentioning Herschel in the first three editions, but he dropped it from the fourth and later editions when he came to suspect that all nebulae were members of the Milky Way, and therefore too close to be used this way.

Herschel's interpretation of the evidence concerning planetary nebulae culminated in his 1791 paper, which was given a title reminiscent of Kant: *On Nebulous Stars, Properly So Called.* We can do no better than to quote from the introductory paragraphs of that paper:

> In one of my late examinations of a space in the heavens . . . I discovered *a star of about the 8th magnitude, surrounded with a faintly luminous atmosphere, of a considerable extent.* The phenomenon was so striking that I could not help reflecting upon the circumstances that attended it, which appeared to me to be of a very instructive nature, and such as may lead to inferences which will throw a considerable light on some points relating to the construction of the heavens.

Herschel then quoted from his observing log concerning sweeps made between 1783 and 1789 in which he discovered peculiar nebulous stars. In regard to one such entry he stated:

> It must appear singular that such an object should not have immediately suggested all the remarks contained in this paper; but about things that appear new we ought not to form opinions too hastily, and my observations on the construction of the heavens were then but

entered upon. In this case, therefore, it was the safest way to lay down a rule not to reason upon the phenomena that might offer themselves, till I should be in possession of a sufficient stock of materials to guide my researches.

Finally the entry of November 13, 1790: "A most singular phenomenon! A star of about the 8th magnitude, with a faint luminous atmosphere, of a circular form. . . . The star is perfectly in the center, and the atmosphere is so diluted, faint, and equal throughout that there can be no surmise of its consisting of stars; nor can there be a doubt of the evident connection between the atmosphere and the star." This object is illustrated on page 115.

The logic of Herschel's argument was persuasive in its simplicity. He recognized two alternative consequences of insisting that the object be composed of stars, and he felt compelled to reject both. If on the one hand the *nebulosity* were composed of ordinary stars that are indistinguishable because of distance, it would follow that the central star must be of an "enormous size" which "outshines all the rest in so superlative a degree as to admit of no comparison." On the other hand, if the *central object* were an ordinary star, how small must be the components of the surrounding nebula! From these considerations he concluded, "We therefore either have a central body which is not a star, or have a star which is involved in shining fluid, of a nature totally unknown to us."

Herschel preferred to assume the central object to be an ordinary star much like the sun and its neighbors. He then proceeded to develop the consequences of admitting the existence of luminous, diffuse matter, and his delight was obvious as he wrote, "What a field of novelty is here opened to our conceptions! We may now explain that very extensive nebulosity, expanded over more than sixty degrees of the heavens, about the constellation of Orion; a luminous matter accounting much better for it than clustering of stars at a distance." (See plate, page 107.) Indeed, the sky never again looked quite the same.

The nebula N.G.C. 1514. "A most singular phenomenon," in the words of Herschel. Discovery of this object, which he called a "planetary nebula," convinced Herschel that some nebulae are truly gaseous and form atmospheres about their stars. (*Lick Observatory, University of California*)

An inkling of his later speculations on evolution is found in the next statement: "If this matter is self-luminous, it seems more fit to produce a star by its condensation than to have come from a star." He had come to believe that stars were being formed as he watched, and his later theories were built to explain the formation of stellar groups from nebulosity rather than from isolated stars.

Herschel was well aware that his conjectures might be considered too speculative and he concluded this paper with a plea that other astronomers, furnished with the necessary equipment, might re-examine the objects from which he had argued and put his theories to the test. Unfortunately, it was Herschel's reward for pioneering that no one else had equipment capable of verifying or disproving his observations. No one else saw the sky as he had; his contemporaries could only accept his observations on faith and listen while he described his conclusions. There were none of the debates, none of the alterations of viewpoint, none of the critical reviews in which the science of the more run-of-the-mill astronomers flourished. He made every effort to lead his readers from one step to the next, dividing an argument into the finest possible stages, but his illustrations were meager, his measurements were imprecise, and perhaps for these reasons he was unable to inspire professional colleagues to follow him into cosmology. It is remarkable that the next attempts to construct large telescopes were those of an amateur, William Parsons. And not until after the application of photography to astronomy (by other amateurs) did larger telescopes appear. One reason for this was the preoccupation of astronomers with the measurement of the distances of nearby stars, a process that did not require large telescopes, and another was the obvious ambiguity—so well shown in the work of Herschel—of investigations of geometrical shapes and arrangements. Astronomers of the nineteenth century preferred to study the motions of planets and moons, because the measurements were so precise and the physical theories had been supplied by Newton, and elaborated on by two centuries of mathematicians.

In 1811 (he was then seventy-three years old) Herschel published an elaborate paper on the relationship of stars and nebulae; his aim was to arrange the objects into a sequence so that they might be "viewed in a new light." As an example of his classifications, consider the titles of five successive sections: "Of Nebulae that are gradually a little brighter in the middle"; "Of Nebulae which are gradually brighter in the

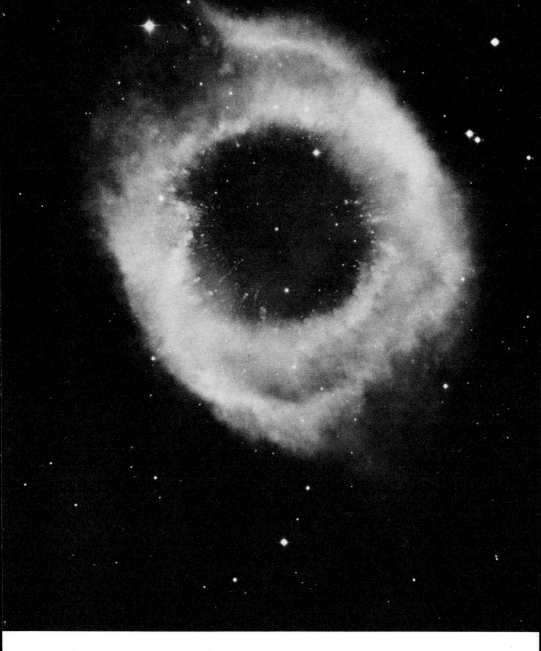

N.G.C. 7293, planetary nebula in the constellation Aquarius. This enormous shell of gas is expanding slowly from the central star. Note the fine streamers and condensations of the interior and the billowy appearance of the outer regions. (*200-inch photograph, Hale Observatories*)

middle"; "Of Nebulae that are gradually much brighter in the middle"; "Of Nebulae that have a Cometic appearance"; "Of Nebulae that are suddenly much brighter in the middle."

His speculations were simple and were contained in two sentences:

> Why should we not look up to the universal gravitation of a matter as the cause of every condensation, accumulation, compression, and concentration of the nebulous matter? Facts are not wanting to prove that such a power has been exerted; and as I shall point out a series of phenomena in the heavens where astronomers may read in legible characters the manifest vestiges of such an exertion, I need not hesitate to proceed in a few additional remarks of the consequences that must arise from the admission of this attractive principle.

Herschel also made a synopsis of that paper, and it provides a handy framework for commenting on the astronomy of his later days and the astronomy of our time. My remarks may cast more light on the shortcomings of his work rather than its strengths, but remember that virtually none of his facts were known before his day and that another century of physics was required to lay the theoretical foundation for his speculation. Notice, too, how much more fertile was his speculative approach than the dry conservatism of William Parsons, the subject of a later chapter. It has been said of Sigmund Freud that he did not know all the answers, but he knew most of the questions. So it was with Herschel.

SYNOPSIS
Astronomical Observations relating to the Construction of the Heavens, arranged for the Purpose of a critical Examination, the Result of which appears to throw some new light upon the Organization of the Celestial Bodies.

"Diffused nebulosity exists in great abundance."
—Herschel increased by a factor of 25 the known nebulosities.

"Its abundance exceeds all imagination."
—In fact, his imagination of it exceeded the reality; in addition to broad, diffuse nebulae associated with stars, such as the one in Orion, Herschel reported others which have not been verified. In reporting thousands of smaller nebulosities, however, he was not mistaken.

"Nebulous matter consists of substances that give out light, which may have many other properties."

—He correctly surmised that many nebulae glow by light emitted by their own gas; not all do, however, as was proven about fifty years later with a spectroscope. The light they emit is the light of nearby stars.

"Either greater depth or greater compression of the nebulous matter may occasion greater brightness."

—In this he was essentially correct, as the total brightness of a nebula is controlled by its total amount of matter, whether diffuse or dense, as well as the brightness of nearby stars which excite it to shine. For example, the Orion Nebula would retain roughly its same brightness if it were more closely compressed about the four bright stars in its center.

"Condensation will best account for greater brightness."

—He meant to say that if a nebula has a bright center as seen in the sky, it is more reasonable to assume a dense center than a spur pointing toward the earth.

"The form of the nebulous matter of round nebulae is globular."

—This is a reasonable assumption, and it is based on the contention that nature does not play tricks. But it is also reasonable to assume that some round nebulae may be spheroids, either elongated or flattened, seen from the particular direction that makes them look round.

"This form is caused by gravitation."

—Again, yes and no. Some nebulae are spherical swarms of stars; they are held together by gravity and they are round because they do not rotate. Other clusters, also held together by gravity, are quite flat because the stars share a common rotation about the center. Our solar system is a fine example of flatness in a system held together by gravity. Newton thought this flatness argued for Divine Creation, but in a later chapter we will permit Laplace to relate his alternative.

"The central brightness of nebulae points out the seat of attraction."

—Herschel discovered that many nebulae show a very pronounced central concentration—a "nucleus." Interestingly, the advent of photography, which tended to emphasize the outer portions of nebulae, led astronomers away from the nuclei. Recent discoveries, based on short exposures, have focused attention on the central condensations again. (See Chapter 21.)

"Progressive condensation may be seen in the formation of nuclei."
—Herschel discovered that if a nebula has a nucleus that is much smaller than the outer portions, the outer portions will be faint relative to the nucleus. This property is now one of the principal means of classifying galaxies.

"It will stop light, and is partly opaque."
—Wrong; luminous gas does not stop starlight in an appreciable amount. Herschel was probably deluded by an optical illusion, or so his description of the Andromeda nebula indicates: "The stars which were scattered over it appear to be behind it, and seem to lose part of their lustre in the passage of their light through the nebulosity." To me, knowing that the stars are in front of the nebula rather than behind it, they do not appear fainter. Herschel rarely, if ever, mentions *dark* interstellar matter, although he did once refer to a "hole in the sky" which is now known to be produced by opaque dust.

"In common good telescopes planetary nebulae cannot be distinguished from stars."

"Perhaps they may in the end be so condensed as actually to become stars."
—No, they are expanding. Anyway, the amount of matter in the nebula wouldn't even start to make a star.

"In thirty-seven years the nebulosity of [the Orion Nebula] *has undergone great changes and much greater since the time of Huygens* [slightly over a century before Herschel]*."*
—Wrong; Herschel's error was a common one, and it resulted from the difficulty of depicting the nebulae faithfully with visual observations and drawings. Only after photography was it proven that most nebulae are very nearly constant over tens of years.

"Nebulae are not permanent celestial bodies."
—But most of them must persist for centuries.

"Conversion of planetary into bright stellar nebulae, into stars with blurs, or stars with haziness."
—Herschel here described objects very much like those Mairan (Chapter 7) had postulated. Today we know little more about their cause than Herschel did.

"When it is doubtful whether an object is a star or a nebula, it may be verified by an increase of magnifying power."

"When the object is very like a star, it becomes difficult to ascertain whether it is a star or a nebula."

—Herschel gave ample evidence that there is no sharp dividing line between the appearance of stellar nebulae and stars. Ironically, the actual difference is enormous, but a very distant nebula (a galaxy in fact) may look precisely like a star except with the most delicate of tests. An expert can notice a softness of the nebular image that would escape a novice.

"When we cannot ascertain whether the doubtful object is a star or a nebula . . . the similitude is as great as any we can expect; for were it greater, there could be no doubt."

—On this delightful note, Herschel ended the synopsis.

Thus Herschel argued from continuity of appearance to continuity of structure—a tool that could cut true but could also cut awry. Most astronomers seem to have reacted in much the manner of Herschel's son John, who wrote four years after his father's death: "So wide is the field of conjecture, and so uncertain the analogies we have to guide us, that we shall do well for the present to dismiss hypothesis, and have recourse (perhaps for centuries to come) to observation."

12 HERSCHEL'S DÉNOUEMENT

In 1792, the First French Republic was established after a bloody revolution; in that year William and Mary Herschel produced John, their only child. William was at the peak of his productivity and the young boy grew up—according to a contemporary, Charles Pritchard— "under the shadow of his father's wonderful telescope." John saw his father and aunt "in silent but ceaseless industry, busied about things which had no apparent concern with the world outside the walls of that well known house."

John grew up to be the epitome of a precocious youngster, responding happily and brilliantly to the stimulation of a philosophical father and a fine English education. He attended Eton for a while, but his health was delicate and his mother wished him back in a small school near home. His father acquiesced and provided him a private tutor in mathematics.

When he was eight, his mother wrote to his doting Aunt Caroline: "Dear Miss Herschel . . . John Herschel has promised to write you before he goes again to school. I thank God he is quite well but rude as ever, he desires me to send his love to you." Another visitor described John as the "little boy entertaining, comical, and promising," clearly implying that John was apt to catch the center of the stage among adults.

In March of 1802, the French and English signed the Treaty of Amiens, and during the brief peace William and his family crossed the English Channel and took a coach to Paris for sight-seeing and conversations with French scientists. Pierre-Simon Laplace, the most illustrious among them, had written to Herschel: "I have learned with great pleasure that you propose to come soon to Paris; I will be delighted to have the honor of knowing you personally, and to tell you enthusiastically of the high esteem your beautiful discoveries have inspired in me."

The sentiment must have been shared by Herschel. Laplace's fame as a mathematician was one of the principal glories of France, and his astronomical work was a fine complement to Herschel's. The two men can be imagined at their first meeting—shaking hands vigorously, smiling broadly, talking happily in praise of each other, neither one really listening to the other, Laplace clapping Herschel on the back, Herschel all the while eager to tell of his latest observations.

John was then ten, and the trip must have offered him one delight after another: meeting Madame Laplace in bed ("which to those who are not used to it appears quite remarkable . . . [She] is a very elegant and well-informed woman," wrote William in his diary of the trip); visiting the Jardin des Plantes and the "place where the Bastille had been"; then going to another garden in the evening.

Napoleon was then First Consul—he had not yet executed his plan for becoming Emperor—and he enjoyed enlightened discussions with scientists. He asked the Minister of the Interior to arrange an audience with Laplace, Herschel, and the eminent physicist Count Rumford, who also happened to be in Paris. The visit merited an elaborate report in Herschel's log. He wrote:

SUNDAY, AUGT. 8 [1802]. . . . About 7 o'clock the Minister con-
ducted Mr. Laplace, Count Rumford and me to Malmaison, the pallas
of the first Consul, in order to introduce me to him. . . . After a considerable walk with [Madame Bonaparte] in the Garden we met with
the first Consul, who was engaged in making improvements in the
garden by directing the persons employed how to conduct water for the
irrigation of the plants. The Minister introduced me and Count Rumford. The first Consul addressed himself to me politely asking some
questions relating to astronomical subjects and after a considerable conversation he also addressed Count Rumford. . . .

The first Consul afterwards entered into a general conversation with
the Minister of the Interior, with Mr. Laplace, Count Rumford and
myself, on common subjects.

After about half an hour's walk he led us towards the house, but
stopping short, on meeting with some other gentlemen, he entered into
conversation with them on the subject of a canal which is to be made
in France. The Consul seemed to be perfectly acquainted with the
subject.

Then the Consul put Herschel on the spot, or so the diary implied:

He now led us to a room where after a short time spent in conversing
he seated himself in a chair, and politely desired me to sit down. As
the same invitation was not given to the rest of the company, nor any
of them took seats, I only bowed my thanks for the first Consul's great
civility and kept standing with the rest. The first Consul then asked a
few questions relating to Astronomy and the construction of the
heavens to which I made such answers as seemed to give him great
satisfaction.

The spotlight swung to Laplace:

He also addressed himself to Mr. Laplace on the same subject, and
held a considerable argument with him in which he differed from
that eminent mathematician. The difference was occasioned by an exclamation of the first Consul, who asked in a tone of exclamation or
admiration (when we were speaking of the extent of the sidereal
heavens): "And who is the author of all this!" Mons. De Laplace
wished to show that a chain of natural causes would account for the
construction and preservation of the wonderful system. This the first
Consul rather opposed. Much may be said on the subject; by joining
the arguments of both we shall be led to "Nature and nature's God."

The conversation turned to some "not particularly interesting" topics, after which ices of "excellent flavour" were served and the group disbanded.

Laplace and Herschel evidently had ample time for serious discussion in the carriage traveling to and from the "pallas." We read:

> As M. La Place and I went and returned in a carriage by ourselves, I led the conversation upon the subject of my last paper, of which I gave him some of the outlines. I mentioned the various possible combinations of revolving stars united in double or treble systems; when I mentioned three stars at an equal distance revolving round a center, he remarked that he had shewn in—I believe—his *Méchanique Céleste*, that six stars could turn round in a ring, about their common center of gravity.

The final entry in Herschel's log appears to be an afterthought. It is a brief description of his visit to Messier, whose list of nebulae had guided Herschel in the early stages of his career. Twenty years before this visit, Messier had fallen to the bottom of a dark shaft, where he had lain for several hours before being discovered. His later years were marked by painful vestiges of that accident, frustration over his inability to find proper financial support, and perhaps by his recollections of the difficulty he had experienced getting himself elected to the Académie des Sciences.

Herschel was sympathetic, but he obviously was not charmed. He wrote:

> A few days ago, I saw Mr. Messier at his lodgings. He complained of having suffered much from his accident of falling into an ice-cellar. He is still very assiduous in observing, and regretted that he had not interest enough to get the windows mended in a kind of tower where his instruments are [Hôtel de Cluny]; but keeps up his spirits. He appeared to be a very sensible man in conversation. Merit is not always rewarded as it ought to be.

In 1813, William's son John graduated from St. John's College at the top of his class and was elected into the Royal Society—surely one of its youngest members—on the basis of an elegant mathematical paper he had submitted the previous year, and a letter of endorsement from his father.

Evidently he had calmed down; a visitor to the household wrote, "He is a prodigy in science and fond of poetry, but very unassuming." Of the father, the visitor wrote:

> Now for the old Astronomer himself; his simplicity, his kindness, his anecdotes, his readiness to explain and make perfectly perspicuous too his own sublime conceptions of the universe are indescribably charming. He is 76, but fresh and stout, and there he sat . . . alternately smiling at a joke, or contentedly sitting without share or notice in the conversation. Any train of conversation he follows implicitly; anything you ask he labours with a sort of boyish earnestness to explain.

John had difficulty choosing a career for himself; only after sorties into mathematics and law, and some important work in chemistry and optics, did he decide to take up astronomy where his father had left off. A recent biographer, Gunther Buttmann, feels that John made a conscious sacrifice when he entered astronomy, and I have the impression that the young man's interests were so diverse that *any* choice would have involved a sacrifice.

What was John's father doing when the choice had to be made? What did John think he was taking up?

At the time of his son's graduation, the "old Astronomer" was putting the final touches on a companion paper to the 1811 discussion of the "Organization of the Celestial Bodies," described in some detail in Chapter 11.

Herschel's 1814 paper was intended to "display the sidereal part of the heavens, and also to show the intimate connection between the two opposite extremes, one of which is the immensity of the widely diffused and seemingly chaotic nebulous matter; and the other, the highly complicated and most artificially constructed globular clusters of compressed stars. The proof of an ultimate connection between these extremes will greatly support the probability of the conversion of the one into the other."

This paper represented the culmination, although not the conclusion, of Herschel's speculation on the nature of the Milky Way. He had adopted a view that is closely related to modern cosmology: the Milky Way has evolved from a rather smooth nebulosity into a fragmented field of stars; some stars are grouped in pairs, and pairs of pairs, others

Herschel as an old man. A visitor to his house wrote of him, "His simplicity, his kindness, his anecdotes, his readiness to explain . . . are indescribably charming."

in clusters. Among the clusters, Herschel found elongated shreds and smooth globes.

He characterized the spherical clusters as "most artificial," but this description is no longer considered appropriate. A blob of gas left alone in space will accumulate into a sphere, and if stars then form they will naturally swarm with the symmetry of their parent. Even in Herschel's day, to call a natural object "artificial" would have seemed quite odd— artificial in itself. The objects are not extraordinarily rare, so although he might have called them unusual, he could not claim that they were the most unusual of the visible objects—the ring-shaped planetary nebulae should have taken that label. I think this appellation has a deeper significance. If we grant that he chose the word purposefully, we may take it as an indication of his attitude toward geometric forms in nature and toward the use of mathematics in science. Herschel implies that chaotic objects were natural objects, while objects that mimicked simple geometrical forms were artificial or unnatural. The contrast of this attitude with the ancient Greek tradition—that nature is modeled after mathematical forms—could not have been more complete, and it is consistent with Herschel's total abstinence from arguments couched in mathematical language. His mathematics rarely rose above arithmetic; his calculations were aimed to evaluate numbers, not to investigate causal relationships. His physical arguments were expressed intuitively, and the fact that they were grounded in Newtonian physics did not save him from wandering off into misconception. His persistent inference of generic relationship from similarity of appearance is the most important example of his failure.

If I seem harsh on William Herschel, imagine that I am using him as a scapegoat to force home the distinction among the uses of mathematics in science. It is clear that he could have employed more formal arguments if he had wished; he could have taught himself the mathematics, but he simply did not think mathematics relevant to his search for the construction of the Milky Way. His intuition was enough— partly, I suspect, because he did not think he had data which merited elaborate mathematical analysis, and partly because he was not enchanted by mathematical reasoning.

What is so enchanting about mathematics? The modern scientist does not feel that the world is *governed* mathematically, but he knows that some of its behavior can be *described* mathematically, and for him this is enough encouragement to take up mathematics.

Herschel did use mathematical reasoning in his last paper on the Milky Way, published in 1817. In some respects this final paper was an anticlimax after his earlier delineation of the Milky Way, his pioneering into celestial evolution, and his suggestion that the Andromeda Nebula was another Milky Way external to our own. Herschel had set aside evolution and he evidently saw little point in iterating the annoying fact that the Andromeda Nebula was now an ambiguous object; he turned to a test of the idea that stars are uniform in brightness and evenly distributed in space. Beyond the obvious importance of the question was an implication for his earlier work, because verification of the simple model would have vindicated his attempts to plumb the periphery of the Milky Way by counting faint stars. He knew that the basic premise, that all stars have the brightness of the sun, was false—his double stars had proved this to him—but he lacked a substitute so he decided to see whether the deviations from this uniformity were severe enough to be detected.

His starting point was a catalogue listing the number of stars of each "apparent magnitude," a system of brightness classification established by the ancient Greeks, who had put the twenty brightest stars in the first magnitude and arranged the remaining visible stars into five equal steps. The Greeks had made no pretense of knowing the actual brightness of the stars—their theories did not require such data—and no one else had succeeded in measuring the stars; so Herschel started from the assumption that second-magnitude stars were twice as far away as those of the first magnitude, that third-magnitude stars were three times as far, etc. This assumption was the child of desperation, but Herschel had no choice—rather, he had a wide variety of choices because he had no facts. He assumed the stars of successive magnitudes to be confined to successive shells of increasing radius but equal thickness, so the calculation of the relative volumes of space allocated to stars of each magnitude was a simple matter. Then, assuming that the number of stars is proportional to the volume of space, he prepared a table of expected numbers within each magnitude class.

In this way he predicted that second-magnitude stars should be 3.8 times as abundant as those of first magnitude, and this result agreed nicely with the actual counts, which gave a ratio of 3.4. But he computed that the fifth-magnitude stars should be 23 times as abundant as those of first magnitude, and this fell far short of the counted ratio of 68. Herschel had not given the faint stars enough room—he had put them too close to the sun.

But there is a flaw in the calculation: some stars are many times as far away as others of the same apparent brightness. Herschel guessed that this must be the case, but he pretended to think it would not make too much difference. In the following comment Herschel admits the truth, and then appears to denigrate it: "The presumptive distances of the stars pointed out by their magnitudes can give us no information of their real situation in space. The statement, however, that one with another the faintest stars are at the greatest distance from us, seems to me so forcible that I believe it may serve for the foundation of an experimental investigation."

This statement of Herschel's is not self-contradictory. He is making a very important distinction between the two aspects of the problem: the assumption that all stars had the same intrinsic brightness; and the use of magnitudes, whose meaning was unclear. He realized that the magnitudes might be the culprits, so he spent his last years attempting to develop new ways of measuring quantitatively the brightnesses corresponding to each magnitude.

His results were reasonably good, but it was not until shortly after his death that astronomers succeeded in accurate measurement of brightness. They then discovered that the steps assigned by the Greeks represented *equal ratios* of brightness, and stars of the sixth magnitude were 100 times fainter than those of the first magnitude—each magnitude giving roughly a factor 2½. Thus, if all stars had the same intrinsic brightness, sixth-magnitude stars would be ten, not six, times as far as those of the first magnitude. (Still later, experimental psychologists discovered that the ratio-stepping of the Greeks was merely one example of a general law that equal differences of impression correspond to equal ratios of stimulus.)

As Michael Hoskin, a biographer of William Herschel, has pointed out, this obstinate insistence that all stars are equal to the sun, so faintness implies great distance, is not sufficiently "unscientific" to bring our disapproval down on Herschel's head. He had come to the idea before he realized that the unequally bright members of a double star are at equal distances from us; he had no reason to think that stars varied by a tremendous amount from one another; he knew that if they differed only by a factor of five or so, his method would still be better than no method at all, at least for certain types of studies.

The concluding note of his attempt to outline the Milky Way was the admission that the "space-penetrating power of the twenty-foot telescope could not fathom the profundity of the Milky Way, and that the stars which were beyond its reach must have been farther away from us than" 900 times the distance of first magnitude stars. He doubted that even his forty-foot telescope would reach the limits of the Milky Way, although he figured it would take him 2300 times the distance of the nearby stars.

The Milky Way is now known to be about 5000 times as large as the distance of the nearby stars, so Herschel was correct in realizing he could not see single stars at the edge—although he would have been gratified to know how close he had come.

If at this point Herschel's conclusions seem vague and shifting, if the relationships of the Milky Way, the clusters, and the nebulous patches are still obscure, then we can sympathize with the perplexed astronomers that William left behind when he died in 1822. Herschel himself was unsure; he became less sure with the passage of time. He wrote in 1811:

> I must freely confess that by continuing my sweeps of the heavens my opinion of the arrangement of the stars and their magnitudes, and some other particulars, has undergone a gradual change. . . .
>
> For instance, an equal scattering of the stars may be admitted in certain calculations; but when we examine the Milky Way, or the closely compressed clusters of stars . . . this supposed equality of scattering must be given up. . . .
>
> We . . . surmised nebulae to be no other than clusters of stars disguised by their very great distance; but a longer experience and a better acquaintance with the nature of the nebulae will not allow a general admission of such a principle.

Despite this confession, most astronomers seem to have assumed that Herschel had clung to his disk theory—stars uniformly scattered within a disk resembling a grindstone. The first statement to the contrary is found in a book by F. G. W. Struve, an influential astronomer who had left Germany to become director of Russia's principal observatory. He wrote, after a description of Herschel's papers:

> We thus come to the results, perhaps unexpected but uncontestable, that the system which Herschel announced in 1785 [the disk theory], collapsed in all its parts, as a result of the later research of its author, and that Herschel, himself, totally abandoned it. . . . The explanation of the Milky Way has rested nearly stationary since the death of Sir W. Herschel. But one might ask why astronomers generally held to the old picture of the Milky Way, announced in 1785, although it had been entirely abandoned by the author himself, as we have demonstrated. I believe that we must look for the explanation in two circumstances.
>
> It was a complete system, imposing because it had a sturdy and precise geometric construction; and the author never revoked it entirely.

Herschel had discovered a new continent in the sky; and he had discovered its enigmas but never resolved them: with courage and grace that are unique in the history of astronomy, he had dismantled his own constructions.

His work is a monument to the human spirit; he truly lives among the stars.

13 THE EVOLUTION OF THE NEBULAR HYPOTHESIS

In the literature of astronomy there are few finer examples of the evolution of a man's thinking than the successive editions of the popular book *The System of the World,* by Pierre-Simon Laplace. Between 1796 and his death in 1827, Laplace prepared six versions of the book, rearranging and amending the material, and deleting ideas that no longer seemed valid to him. Here we see evidence of the growth (and the occasional retrenchment) of astronomy during the latter part of Herschel's career, and we find evidence that Herschel's discoveries profoundly influenced Laplace's theories of the origin of the solar system.

Laplace was born in 1749, the son of a farmer in southern Normandy. He soon showed outstanding mathematical ability and a prodigious memory. At eighteen he was teaching mathematics at the

Military School in his home town but craved the more stimulating atmosphere of Paris. He therefore collected a few letters of recommendation and took them to Jean Le Rond d'Alembert, the most prominent mathematician in France. He received no reply, so he prepared his own recommendation: a short mathematical treatise on mechanics. This was delivered to d'Alembert, and within a day Laplace was summoned for an audience. The immediate result was a professorship in the Military School in Paris; Laplace's self-confidence had been justified.

The young man soon became a giant in the mathematical world. Displaying extraordinary intellect, he expected as much of his readers as of himself: his papers often left gaps in the argument, saying *"Il est facile à voir"* when in fact many did not find it so "easy to see." Laplace's attitude toward mathematics was similar to that of Newton, his predecessor by nearly a century. For both men the goal was to understand the physical world, and mathematics was adopted as the tool; neither man took great delight in elegant mathematics for its own sake.

Newton had created the foundations of calculus in order to investigate the motions of the planets, and in the century that followed, many of the ablest mathematical minds had continued to build on those foundations. Laplace therefore found an elaborate structure awaiting him.

One of his earliest triumphs was to resolve an enigma that Newton had placed in the hands of the Almighty; namely, the endurance of the solar system. Newton had suspected that the present arrangement of the solar system—planets circling the sun in nicely separated orbits—could not endure, because each planet was gravitationally attracted by all its neighbors, and the accumulation of these attractions, no matter how delicate they might be, would ultimately bring disaster. Newton did not have the mathematical machinery needed to crack this problem, but he supposed that disaster had been averted by the intervention of the Almighty, who set things right from time to time.

Shortly after, Newton's view seemed to be confirmed: Jupiter's orbit was found to be shrinking and Saturn's orbit was expanding. No one knew where it would end, but astronomers supposed that the solar system was disintegrating. Laplace showed that the danger was ephemeral; he proved that the behavior of the planets was *periodic*. They would reverse themselves every 929 years, and therefore were not being carried to destruction.

It is impossible to know what Laplace thought of this accomplishment, but he could not have been unaffected by it. He had proven that the continued existence of the solar system did not require Divine repairs; the machinery of the universe was so smoothly built that it could operate without assistance. This idea was not totally new—in fact, the Greek conception of the universe as an assembly of crystalline spheres which rotated without assistance was rather similar—but the context was new, and the implications were far-reaching. The Ancients gave their gods other activities: they operated storms and battles, and other affairs closer to home; the sky was merely the arena of the earth, a backdrop for the sun. But by the time of Laplace, the planets and stars were recognized as extensions of our own world, and therefore it made no sense to imagine that God ruled only our own planet. The solar system and the stars were His seat also, and Laplace was removing Him from His most splendid throne, the assembly of the planets. A story forming part of the Laplace legend relates that when Napoleon Bonaparte asked Laplace whether he had left any place in the universe for the Creator, Laplace said: "Citizen, premier consul, I have had no need of such an hypothesis."

I would take this as confirmation of Laplace's atheism, but another writer of the twentieth century has said: "If that was the response really made, I do not see at all the ground for the irreverential or atheistical attitude often attributed to Laplace. There may indeed be a very deeply religious sentiment in the belief of a universe so harmoniously constituted that there would be no need of continual 'retouching' for it to preserve its course."

By the time he was fifty, Laplace had become involved in politics; he served as Minister of the Interior for Napoleon in 1799, but he proved to be an abysmal administrator. After six weeks Napoleon removed him, placed his own brother in the office, and gave Laplace a seat in the Senate as a consolation. In his *Memoires of Saint Helena*, Napoleon treated the incident with irony: "Geometer of the first rank, Laplace lost no time in proving himself to be a worse than mediocre administrator; from the very first, we saw that we had made a mistake. Laplace never seized a question from its proper point of view: he sought subtleties everywhere, saw only the problems, and finally brought the spirit of the infinitesimal into administration."

In his portrait, Laplace has the mien of an aristocrat: a lengthy nose and a sharp chin are separated by a pair of fine lips. His smile suggests toleration more than pleasure. Laplace was an aristocrat of the intellect; he saw no reason to imagine limits to the knowable. Boundaries on the known were merely temporary and they would be pushed back ever further by the application of man's finest skills. To *know* was for him the highest achievement of man.

This thirst for mathematical and physical knowledge controlled his life; he worked at science through four stages of French history. Commencing with the Revolution, he continued through the Republic, the Empire, and the Restoration, and he invariably dedicated his newest work to the current rulers and seemed to stand in their favor. One biographer used the word "supple" to describe his political manner, and a later biographer says that is putting it kindly.

No doubt these opinions seem to be supported by the fact that Laplace was among the first to vote for the dismantling of the Empire and the removal of Napoleon, despite the honors and benefits conferred on him by the Emperor. But I cannot avoid the impression that Laplace's gratitude for those favors must have been tarnished by Napoleon's obvious desire to add a few bright stars to his own constellation. Royalty has always done so; scholars have often responded gratefully, but I doubt that either party felt that the liaison was to become a marriage. In any case, Laplace's first and only loyalty was to Science.

A portion of Laplace's popular book *The System of the World* dealt with the origin and development of the solar system. This was not the first such discussion but it was by far the most influential, and for almost two centuries the ideas of Laplace held the center of the stage. The influence of the book was due in part to the reputation of its author. It was also partially due to the systematic manner of the exposition. Finally, it was due to Laplace's exploitation of observational evidence.

Laplace's mathematical investigations of the *future* of the solar system, aimed at determining whether the planets would remain in their present orbits, naturally led him to consider its past—and the first two editions of the book (1796, 1799) contained brief comments on the origin of the peculiar fact that all the planets follow orbits that are nearly circular and that lie nearly in the same plane. When Laplace wrote those editions the relevance of Herschel's work to his own would have been far from obvious. In the third edition, published in 1808 (after Herschel's visit), the comments were greatly expanded, and in the fourth (1813) he enlarged them still further and set them into a separate section, known as "Note VII" of the Appendix.

Herschel's name appears for the first time in the fourth edition, when Laplace had finally realized how nicely Herschel's observations seemed to corroborate his own speculations.

The following version of the final chapter of *The System of the World* has been constructed from various editions, and the contributions of each edition have been marked with a different type face. As presented here, the text does not correspond to any of the actual editions, but it does display the structure of Laplace's theory and virtually all of its salient points. A chronology of the publication precedes the text.

KEY TO TYPE FACES

1. Text in roman appeared in the first edition.
2. *Text in italic was inserted in the third, unless another edition is specified.*
3. (Roman text in parentheses appeared only in the fifth and sixth.)

CHRONOLOGY OF *THE SYSTEM OF THE WORLD*

French Republic established; Herschel announces nebulosity	1792
First edition	1796
Second edition	1799
William Herschel visits Paris	1802
Herschel elaborates his ideas on nebulae and stars	1803
Third edition	1808
Fourth edition	1813
Herschel's final publications	1814, 1817
Fifth edition	1824
Death of Laplace	1827
Sixth edition (Laplace had reviewed the proofs)	1835

THE SYSTEM OF THE WORLD
PIERRE-SIMON LAPLACE

Consideration on the System of the World on the Future Progress of Astronomy

(The history of astronomy presents three distinct periods which are concerned successively with the phenomena, the laws governing them, and the forces on which the laws depend; such has been the course followed by this science, a course that the other natural sciences ought to follow. The first period covers the observations of apparent celestial motions before Copernicus and the hypotheses that were imagined to explain them and make them amenable to computation. In the second period Copernicus deduced the rotation of the earth on itself and about the sun, and Kepler discovered the laws of planetary motion. Finally, in the third period, Newton, starting from these laws, advanced the law of universal gravitation; geometers then applied this principle and derived from it all the phenomena of astronomy and the numerous irregularities of the motions of planets, satellites and comets. Thus, astronomy became the solution to a great problem of mechanics. . . . It has all the certainty that follows from the immense number and variety of rigorously explained phenomena and from the simplicity of the principle used to explain them. Far from fearing that a new star or planet will contradict this principle, we can affirm in advance that its motion will conform to it: such we have seen for ourselves regarding Uranus and the recently discovered asteroids; and each apparition of a comet provides new proof.)

The addition of this historical introduction to the fifth edition reflects Laplace's realization that the laws of Newton and the concept of universal gravitation had been extended—largely by the work of Herschel —to planets invisible to the naked eye, and to stars rotating in couples well beyond the confines of the solar system. Herschel's evidence convinced astronomers that they had the tools for investigating the far reaches of the universe.

Let us cast our attention on the arrangement of the solar system and its relation to the stars. The immense globe of the sun, focus of the principal planetary motions, turns on itself in twenty-five and a half days; its surface is covered with an ocean of luminous material whose lively effervescence forms varying spots, often quite numerous and sometimes larger than the earth.

Laplace refers to the solar atmosphere as an "ocean," and he meant the word to be taken quite literally. At that time, the sun was thought to be solid in its interior and liquid in its outer layers. Not until the end of the nineteenth century could physicists describe its composition with an air of conviction: the sun was then proven to be gaseous throughout, the center being too hot to remain solid.

> Above this ocean rises a vast atmosphere; it is here that the planets with their satellites move in nearly circular orbits which are inclined very little to the solar equator. Countless comets, after approaching the sun, recede to distances which prove that the sun's domain extends far beyond the known boundaries of the system of planets. Not only does this star act on these bodies and force them to move around it, it spreads on them its light and heat. Its benevolent action permits animals and plants to spring up and cover the surface of the earth, and analogy leads us to believe that it produces similar effects on the planets; because it is not natural to suppose that the material whose fecundity we see develop in so many ways would be sterile on so large a planet as Jupiter, which, like the earth, has its days, nights and years, and where changes implying very active forces have been observed. Man, made for the temperature which he enjoys on earth, apparently could not live on other planets; but ought there not be an infinite variety of configurations suitable to the varied temperatures of celestial bodies? If the mere difference of composition and climate gives such variety in earthly objects, how much more must those of the various planets and their satellites differ? The most active imagination cannot conceive them, but their existence is most likely.
>
> Although the geometrical elements of the planetary system are physically independent of each other, there are, nevertheless, certain relationships among them which can clarify their origin. On close consideration, it is astonishing to find all the planets moving about the sun from west to east and almost in the same plane; all of the satellites move about their planets in the same sense and nearly in the same plane as the planet; finally, the sun, the planets and their satellites, whose rotary motion we can observe, turn on themselves in the direction and nearly in the plane of their orbital motion.

The remarkable arrangement of the solar system was the starting point of Laplace's cosmogony, as it had been for Swedenborg and Kant. Newton had remarked on this arrangement but, failing to see how it could have arisen from forces now acting among the planets, he had turned his back on the problem rather than entering into speculation. He laid the problem in the lap of the Almighty, and preferred to think of this order as a sign of the power of God. Laplace looked for other forces to align the planets in their orbits. Emboldened by the conviction that he was in communication with the Divine, Swedenborg had combined Newton's piety with Laplace's intellectual courage, and he had speculated with seeming abandon.

But just how remarkable is the arrangement of the planets? Why was Laplace so sure that it required a special explanation? Laplace's argument is the following:

> Such an extraordinary phenomenon can hardly have haphazard causes; it suggests that a general cause has established all of the motions. To obtain an estimate of the probability of such a cause, we remark that the planetary system as it is known today, comprises seven (*eleven*) and fourteen (*eighteen*) satellites; we have observed the rotation of the sun, of five (*six*) planets, of the moon, *the satellites of Jupiter,* the ring of Saturn and one of his moons. These form an ensemble of thirty (*forty-three*) motions directed in the same sense. . . .
>
> *The chances are four million to one that this arrangement is not haphazard; this is a likelihood far greater than the most certain events of history, about which we permit no doubts. Thus we must believe, with at least that much confidence, that a primitive cause has directed the motions of the planets, particularly if we consider that the inclinations of most of the motions are much less than a quarter circle.*

Laplace's calculation was based on the following idea:

If we see a string of beads lying on a table we are not astonished to see them forming a circle, or an oval, say, because we know they are held in position by the string running through them. If, however, we were to discover that there was no string confining the beads—in other words that the beads had the complete run of the table—we might be more intrigued. We would probably assume that they had been carefully aligned. In just this way Laplace assumed that although the planetary orbits have the complete run of tilts and eccentricities (elongations) available to them, they have been confined to a very small set of the possibilities.

He said the certainty that they were *not* simply thrown together is four million to one, and he claimed this to be more certain than what we read in history books. In later editions Laplace raised the certainty by a factor fifty: more satellites and telescopic planets had been discovered with the same regularity of pattern.

The calculation is a valid reason for being astonished at the form of the solar system, but Laplace's reference to the certainty of what we read in the history books might be considered a joke if Laplace had shown other signs of humor in his writing. Discussions of the probability of history have been the cause of much heat but little light, because no one has yet been able to outline *a priori* limitations to alternative historical possibilities.

> An equally remarkable phenomenon of the solar system is that the orbits of the planets are nearly circular, while those of the comets are highly elongated; the orbits of this system offer no intermediate nuances. Again we are compelled to recognize the effect of a regular cause; happenstance alone could not possibly give an almost circular form to the orbits of all the planets. Whatever arranged these orbits also made them nearly circular. Moreover, the same cause must explain the great elongation of cometary orbits, and the fact that comets move in all directions as though they had been thrown at random.
>
> Thus to trace back to the cause of the original motions of the planetary system, we have the following five facts: the motions of the planets in the same direction, and in almost the same plane; the motions of the satellites in the same direction as that of the planets; the rotary motions of these various bodies, and of the sun, in the same directions as their trajectories about the sun and in approximately the same planes; the nearly circular orbits of the planets and satellites; finally, the great elongation of the orbits of the comets, although their orientations have been left to chance.
>
> Buffon is the only one I know who has attempted, since the discovery of the true system of the world, to trace the origin of the solar system. He supposes that a comet, by falling into the sun, expelled a torrent of material which reunited into globes of various sizes and at various distances; these globes, after cooling and becoming opaque and solid, are the planets and their satellites.

Buffon's theory that the planets had been torn from the sun by a passing body had been published shortly before Laplace's time in a widely read account of natural history, so it is not surprising that *this* should have been the theory familiar to Laplace when he wrote the first edition of his *System of the World*. That he should have remained unaware of the speculations of Kant right up to the time of the sixth edition, however, makes him seem a bit myopic.

> This hypothesis fits the first of the five phenomena listed previously, because it is clear that all the bodies thus formed will move near to a plane passing through both the center of the sun and the torrent of

Pierre-Simon Laplace's Nebular Hypothesis. According to this theory, the sun contracted from an immense rotating cloud. Successive rings of matter were imagined to detach during the contraction and to form the planets. (*Reproduced from* Physics Today, *Vol. I, No. 6* [*1948*], *p. 14*)

material which produced them; the other four phenomena appear inexplicable to me by this hypothesis. In fact . . . the direction of rotation of the planets in this hypothesis is not necessarily the same as that of the orbital motion . . . and this difficulty applies to the rotation of the satellites as well. . . .

The circularity of the planetary orbits is not only very difficult to explain with this hypothesis, the facts seem to contradict it. If a body moving in an orbit about the sun comes close to the surface of this star, it will return unfailingly at each revolution; from this it follows that if the planets had originally been detached from the sun they would touch it again at each return, and their orbits would be far from circular. . . .

Finally, there is no evident reason in the hypothesis of Buffon why the eighty (*one hundred*) comets observed thus far should have such elongated orbits. This hypothesis is far from fitting the facts. Let us see whether it is possible to deduce their true cause.

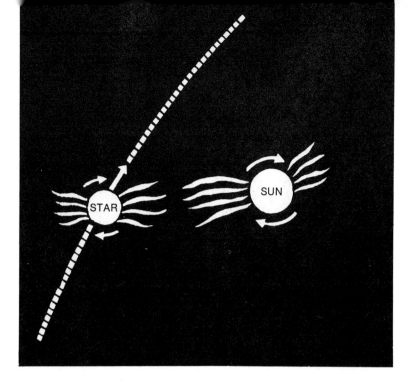

The Tidal Theory. An alternative theory for the formation of the planets supposes that a passing star raised great tides on the sun and released the gas which subsequently formed the planets. It no longer appears likely that this process will produce a planetary system. (*Reproduced from* Physics Today, *Vol. I, No. 6* [*1948*], *p. 14*)

Having eliminated the competition, Laplace proceeds to develop his own hypothesis: The sun once had an enormously extended envelope. Laplace suggests that the nova discovered by Tycho (see Chapter 3) in 1572 might be an example of a distended star, and this idea is similar to the current theory that a nova outburst is an explosion that lifts the outer layers from a star.

> Whatever the sun's nature, it must have encompassed all of the planets; and considering the enormous distances separating these bodies, it must have been a fluid of an immense extent. In order to have given the planets almost circular motions in the same direction, this fluid must have surrounded the sun like an atmosphere. The consideration of the planetary motions thus leads us to think that, by virtue of an excessive heat, the solar atmosphere originally extended beyond the orbits of all the planets and that it progressively shrank

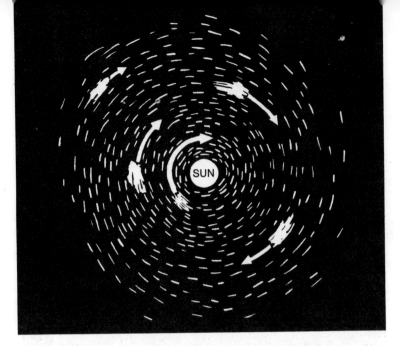

The Planetesimal Hypothesis. At the close of the nineteenth century, astronomers came to feel that Laplace's Nebular Hypothesis could not explain how the matter of the sun could be released in rings and, if it were, how the rings would form into planets. T. Chamberlain and F. C. Moulton suggested that matter was released by the tidal action of a passing star and that it then cooled into small particles, which aggregated into the planets. Recent theories retain Laplace's idea of a single nebula, but they include the Chamberlain-Moulton idea of cooling while still in an extended, turbulent cloud. (*Reproduced from* Physics Today, *Vol. I, No. 6* [*1948*], *p. 14*)

to its present limits. This might have occurred through causes similar to those which made the famous star of 1572 suddenly shine so brightly for several months in the constellation of Cassiopeia.*

The great eccentricity of cometary orbits leads to the same result; it suggests the disappearance of a large number of comets with less-eccentric orbits, as though they had been destroyed by passing through an atmosphere. If this were the case, the only comets existing today would be those which were outside the atmosphere during its enormous extension; and, as we can only observe comets which come quite close to the sun, the observable comets would now be those with very

* This speculation was omitted from the fourth and successive editions, although in the fifth and sixth editions, Laplace listed the star of 1572 as an example of variability in the heavens.

eccentric orbits. And, at the same time, we can see that the orientation of their orbits would be scattered at random, because the solar atmosphere could not have influenced them.

Laplace's ideas about comets changed drastically in the fourth edition. Whereas in earlier editions he said that the comets might represent debris from the early days of the solar system, the fourth edition presents the suggestion that the comets are strangers to the solar system. He supposed that they were small nebulae of the type seen by Herschel and that they had wandered among the stars until captured by the sun. These theories were eliminated from the fifth and sixth editions, but they are remarkably similar to views developed by Jan Oort and Fred Whipple in the last few decades: Pre-planetary fragments of ice and dirt, left at the periphery of the sun's sphere of influence, are deflected into the neighborhood of the sun by the passage of nearby stars; the elongation of comet orbits would then be a natural consequence of the great distance of their birth-sites.

The first explicit mention of Herschel's discoveries of stars embedded in nebulosity appears in the fourth edition, where Laplace gives a lengthy review of Herschel's reasons for thinking that such nebulosity is actually condensing on the surface of the star. Laplace points out that the natural extension of this idea is that stars should appear in clusters if a single nebula can produce many stars. Modern astronomers agree that the observed clusters were formed from large clouds of dust and gas; they also find that many of the clusters must be relatively young, because the force of fluctuating circumstances and the swirling rotation of the galaxy tend to disrupt them.

However, it is not now certain that the nebulae described by Herschel are stars in the process of formation. For example, the gas is *expanding* from the planetaries rather than contracting, so the star is shedding matter rather than accumulating it.

Recently, groups of small red stars buried in thick nebulosity have been discovered, and the current view is that they may represent the infantile stars postulated by Laplace. (See Chapter 3.)

During* the primitive state we suppose for the sun, it resembled

---

* The following four italicized paragraphs appeared in the fourth and successive editions; Laplace rearranged them several times.

the nebulae which the telescope shows to be composed of a nucleus surrounded by a nebulosity. By condensing on the surface of the nucleus, such a nebulosity is being transformed into a star. By extension, one can imagine an earlier state, preceded itself by other states, in which the nucleus was more diffuse and less luminous. Thus, one arrives, by tracing back as far as possible, at a nebulosity so diffused that its existence would only be suspected with difficulty.

Such, indeed, is the initial state of the nebulae which Herschel has observed with particular care by means of his powerful telescopes. He has followed the progressive condensation, not within a single nebula—centuries would be required for its detection—but within the ensemble, in much the way that we may follow the growth of trees within a forest by examining individuals of different ages. Nebulosity is spread over a large extent of the sky, and Herschel found it principally among clusters of stars. In some clusters, he found it weakly condensed about several faint nuclei; in others, the nuclei were much brighter than the surrounding nebulosity. When the atmospheres of the nuclei have separated by a further condensation, the result is a multiple nebulosity formed of bright, tightly packed, nuclei. Occasionally, the nebulous matter condenses uniformly into nebulae named planetary. Finally, a more intense condensation transforms all of these nebulae into stars. Classed according to this philosophy, nebulae convincingly declare their future transformation into stars. The following considerations provide tests of these ideas.

For some time, the unusual arrangement of some of the naked-eye stars has struck philosophical observers. Michell has remarked how unlikely it is that the stars of the Pleiades, for example, should be compressed purely by chance into the narrow space allotted to them.* He concluded that such groups of stars are the results of primitive causes and reflect a general law of nature. These groups are the necessary result of the condensation of nebulae with multiple nuclei, because it is clear that such material will be attracted unceasingly toward the separate nuclei and ultimately form a group of stars like the Pleiades. The condensation toward two nuclei will form two very close stars turning about one another, such as the double stars in which the relative motions have recently been discovered.†

Thus, by the progressive condensation of nebulous matter, one is led downward to a view of the sun as surrounded formerly by a vast atmosphere, a view to which one may also ascend, as we have seen by examining the phenomena of the solar system. Such a remarkable

---

* A similar cluster of stars, containing perhaps a thousand members, is illustrated in the plate on page 39.

† This is an allusion to Herschel's announcement, in 1803, that he had found double stars to be in motion about each other.

encounter along opposite routes gives to this early state of the sun a likelihood approaching certitude. . . .

But how did the extended atmosphere of the sun produce the rotations of the planets? As these bodies would have fallen into the sun if they had penetrated the atmosphere, we may conjecture that they were formed at the successive limits of the atmosphere, in zones of material abandoned as it contracted toward the surface of the sun. We may further conjecture that the satellites were formed in a similar fashion from the atmospheres of the planets. . . . *Consider now the zones of vapor successively left behind. In all likelihood, these zones will condense, as a result of the mutual attraction of their molecules, into rings of vapor circulating about the sun. Collisions among the molecules of each ring will accelerate some and retard others, until all will attain the same angular motion about the sun. Thus, the real velocities of the most distant molecules will be the greatest. . . . If all of the molecules of a ring collect uniformly, they will produce a liquid or solid ring; but the regularity which this formation requires in all parts of the ring will render it an extremely rare phenomenon. For this reason, the solar system offers but a single example, that of the rings of Saturn. Almost always, each ring of vapor will be disrupted into several masses which will circle the sun . . . with a rotation in the same direction as their orbital motion. But if one of them is large enough to attract the others, the ring will be transformed into a single body. . . .*

The five phenomena of the solar system, mentioned earlier, flow naturally from these hypotheses, to which the rings of Saturn add a further degree of likelihood.

The crux of Laplace's hypothesis is that the contracting sun sheds rings of gas ("liquid or solid" in the third edition alone) which condense to form the planets; similarly the contracting planets shed matter which forms their moons. The similarity to Swedenborg's hypothesis is striking, but the differences are more important. Swedenborg had suggested that the planets were ejected from a mature sun as a consequence of its *increasing* size and rotation; from the standpoint of an astronomer this is an entirely different suggestion.

Neither hypothesis fares well in the twentieth century, because each leaves unanswered the difficult question "How can the solar matter accumulate into planets despite its very strong tendency to expand and then settle again into the sun as a diffuse cloud?"

In 1900, F. R. Moulton and T. C. Chamberlain, mathematicians, studied the contraction of a rotating gas cloud and they found Laplace's suggestion to be fraught with difficulties, although—as Moulton admitted—to disprove the hypothesis is not an easy matter, because it requires proving "what cannot happen in any time, no matter how long."

They listed four main objections: (1) Matter will not be released in discrete rings; it will leave the equator of the sun in a steady stream. (2) Even supposing a ring to be formed, the gravitational force of the main body of the sun would disrupt it and prevent its agglomerating. (3) Some satellites in the solar system move against the common flock, and they cannot be explained by Laplace's hypothesis. (4) The marked concentration of planetary matter in Jupiter and Saturn suggests that the solar nebula must have been highly irregular, and this is not consistent with the assumption that it has gradually contracted from a much larger globe of gas.

As an alternative, Moulton and Chamberlain suggested an idea reminiscent of Buffon's theory: A passing star draws a streamer of gas from the sun, whips it around into a spiral nebula, and the nebula then fragments and condenses into separate blobs; these cool and become planets. This model would seem to suffer from the weakness pointed out by Laplace in Buffon's, but the authors stress that smaller fragments might cool and solidify, and then coalesce through collisions.

Recent theories seem to combine attributes of all of these older ideas in various mixtures. But in the final analysis, cosmogonies, like novels, inspire visions of "truth" primarily in the eyes of their authors; most others seem to believe that they are fictions.

> Whatever may be the fate of this theory, which I present with the diffidence appropriate to what is not the result of observation or calculation, it is certain that the arrangement of the solar system is such as to assure its greatest stability—supposing no trouble from external forces. To that end, the motions of the planets and the satellites are almost circular and are directed in the same sense and nearly in a common plane. This system will merely oscillate by a small amount about its average state. It seems that nature disposed everything in the sky to assure the stability of the planetary system, by a design similar to what it follows so wonderfully on earth for the preservation of humans and for the perpetuity of species.

This was written three decades before the initial announcement by Charles Darwin and Alfred Wallace of their theories describing the tendency for species to evolve, and it suggests that in the mind of Laplace the problem presented by living forms is their *stability* rather than their *variation*. Laplace evidently believed that species would deviate from their original patterns if it were not for the restraining hand of the Divine. What particular evidence had convinced Laplace of the "perpetuity of species" is not clear, but it would probably not be far off the mark to assume that he took the common viewpoint of his day: The earth is undergoing violent upheaval; despite this, man stands in much the form he did many thousands of years ago when he was created; therefore he must receive special protection from the Divine.

William Herschel had for the first time described evidence for the gradual evolution of astronomical objects, and it is tempting to suppose that the conception of evolving species as expressed by Laplace in the preceding paragraph can be attributed to the influence of Herschel's discoveries—or that it was sharpened by them. But against this interpretation stands Herschel's own analogy: he compared his sequences of astronomical forms to the sequences of forms that may be discovered in a garden. He commented that young plants and old plants of a particular variety are seen simultaneously, and from a study of their relationships it would be possible to trace the life history of the plants. In other words, it was not the evolution of *species* but the evolution of *individuals* that Herschel thought he was seeing in the sky. In this, most astronomers would concur; the idea of species has to date played virtually no role in astronomy: galaxies have appeared to remain galaxies and stars have appeared to remain stars.

Laplace next launches into a kindly remonstration of Newton and expounds what has become known as the "modern scientific" attitude toward God. In effect, he says that Newton demeans the conception of God by insisting that God is merely the engineer of the universe.

> If my conjectures on the origin of the planetary system are correct, the stability of this system is a direct consequence of the laws of mechanics. The successful explanation of these phenomena and of several others allows us to think that everything depends on these laws through relations that are more or less hidden, but about which it is wiser to confess ignorance than to substitute imaginary causes.*
>
> I cannot refrain from observing how far Newton deviated on this

* In the fifth edition, Laplace added a phrase so the sentence ended: *"imaginary causes conceived solely to calm our uneasiness on the origin of things interesting to us."*

point, from the method he otherwise applied so happily. After exposing at the conclusion of his PRINCIPLES OF NATURAL PHILOSOPHY* *the singular phenomenon of the motions of the planets (being in the same direction, in the same plane, and nearly circular) he adds: "These regular motions cannot have a mechanical cause. . . ."*

But isn't it possible that even this arrangement of the planets is an effect of the laws of physics? Couldn't the Supreme Intelligence invoked by Newton have made the arrangement depend on a more general phenomenon? Such, according to us, would be the diffusion of nebulous matter within clusters and through the immensity of space.

Can we still affirm that the conservation of the planetary system was one of the aims of the author of nature? As Newton supposed, the mutual attraction of the planets cannot alter the stability of the system; but even if there were in space no fluid other than light, its resistance and the decrease that its emission produces in the mass of the sun would eventually destroy the arrangement of the planets. To maintain it, a reform would doubtless be required. But, isn't it true that the many species of extinct animals, which Cuvier so sagaciously recognized among the fossil bones he described, indicate a tendency to change even the most stable things in nature? The size and importance of the solar system must not make it an exception to this general law— its grandeur is merely relative to us, and, huge as it seems, it is only an imperceptible point in the universe. If we go over the history of the human spirit and its errors, we will repeatedly find final causes being pushed back to the limits of knowledge. The causes that Newton pushed to the limits of the solar system were, not long before, placed in the atmosphere to explain the meteors. In the eyes of the philosophers, they are merely the expressions of ignorance whose true cause resides in us.

(Leibnitz, in his quarrel with Newton on the invention of the infinitesimal calculus, sharply criticized the concept of divine intervention to put the solar system back in order. "This is," he says, "to have rather narrow ideas of God's wisdom and power." Newton responded with an equally sharp criticism of Leibnitz's pre-established Harmony, which he called a perpetual miracle. Posterity has not admitted these vain hypotheses, but it has given the most complete justice to the mathematical works of these two great geniuses. The discovery of universal gravitation, and the efforts of its author to thus connect

---

* In the fifth edition, Laplace here added a footnote: *"This remark does not figure in the First Edition of the* PRINCIPLES. *Until then Newton devoted himself only to mathematics, which, unhappily for her and for her glory, he abandoned too early."*

celestial phenomena will always be the objects of admiration and gratitude.)

Laplace's discussion of Newton's attitude toward the Almighty clearly reveals his own mechanistic attitude—an attitude that had been germinated by Newton's successes and which flowered during the eighteenth and nineteenth centuries. According to this conception, the world is a machine, and its entire future may be predicted if we make sufficient measurements today and perform sufficient computations. (To judge from some of his political speeches, Laplace also thought that society itself could be considered a vast machine.)

Now let us cast our eyes beyond the solar system. Numberless suns are scattered through the immensity of space at such great distances from us that the entire orbit of the earth, observed from their positions, would be imperceptible. *Analogy leads us to believe that they are also the foci of planetary systems, and our theory suggests that the analogy would be valid. If these stars were endowed with rotation, like the sun, and were surrounded by a vast atmosphere, it is natural for us to attribute the same results to the condensation of these atmospheres.* Several stars display remarkable periodic variations in their color and brightness: *These indicate the existence of great spots on their surface and rotational motions which alternately reveal and conceal them to our view.*

Concerning variations of stellar brightness and color, astronomers no longer agree with Laplace that spots on the surface of a star will account for most of the observations. Many stars undergo periodic swelling and shrinking; this fact is directly proven by variations of the velocity of the atmospheric gases. Pairs of stars also show periodic variations if the orbits are arranged so that they alternately block each other's light.

. . . Other stars have appeared quite suddenly and then have disappeared after several months. (One example of this was the star observed by Tycho Brahe in 1572, in the constellation Cassiopeia. It quickly surpassed the brightest stars, even Jupiter, and it was seen in the daytime. Its light then decreased and it vanished sixteen months after its discovery. Its color varied widely: it was first dazzling white, then reddish yellow and finally a leaden white like Saturn.) What amazing changes must have occurred on these great bodies to be observed from our distance. Think, how they must surpass what we see on the surface of the sun, and what convincing proof they give that nature is not everywhere and always the same. All such stars which again became invisible remained in precisely the same position during their apparition; thus, there exist large bodies in space, perhaps as numerous as the stars.

A luminous star of the density of the earth, and having a diameter two hundred and fifty times that of the sun would not, by virtue of its gravitational attraction, let any of its light escape to us. Thus, it is possible that the largest luminous bodies of the universe are, for that very reason, invisible.

Laplace's statement, that a monstrous star with the density of the earth and 250 times the diameter of the sun would not be seen, is in itself a monstrosity—no star like that could be formed of matter as we know it.

The statement was removed from the third edition, and it did not reappear. Interestingly enough, the General Theory of Relativity does imply that bodies might be invisible as a consequence of gravity, although they would have a much higher density than Laplace's star.

It appears that the stars, far from being scattered at equal distances, are collected in various groups containing, *in some cases,* billions of stars. *This is another consequence of their origin, according to our hypothesis.** Our sun and the brightest stars are probably part of one such group, which seems to surround the sky and form the Milky Way. The large number of stars seen in the field of a powerful telescope aimed at the Milky Way gives proof of its immense depth, which is more than a thousand times the distance to Sirius. *Thus it is likely that the light emitted by most stars took many centuries to reach us.*

An observer receding from the Milky Way would ultimately see a white and continuous light *of small diameter* because irradiation, which persists even in the best telescopes, would cover and hide the gaps between the stars; it is thus likely that the starless nebulae are groups of stars seen from afar, and that they would give the appearance of the Milky Way if only we were to draw near them. [In the third edition, this sentence had been changed to the more cautious: "it is thus likely that the nebulae are, *for the most part,* groups of stars seen from afar, and that they would give the appearance of the Milky Way if only we were to draw near them." In the fourth edition, the sentence took its final form: *"it is thus probable that, among the nebulae, some are*

---

* This sentence appeared only in the fourth edition. Its deletion is probably a symptom of the uncertainties expressed in Herschel's later papers.

groups which comprise a large number of stars and which, seen from their interior, would resemble the Milky Way."]

The mutual distances of the stars forming each group are at least one hundred thousand times greater than the distance from the sun to the earth; from the multitude of stars in the Milky Way, one can judge the prodigious extent of these groups.* If one now reflects on the immense intervals separating them, the imagination is stunned by the grandeur of the universe and can hardly conceive it to be limited.

From these considerations, based on telescopic observations, it follows that the nebulae which appear sharply outlined and whose centers can be located with precision are, from our standpoint, the most fixed of the celestial objects; they are the objects to which the positions of the stars might be related.*

. . . Seen in its entirety, astronomy is the handsomest monument to the human spirit, the most noble sign of man's intelligence. Seduced by the illusions of his senses and by self-esteem, man has long regarded himself as the center of celestial motion, and his vain pride has been punished by fears inspired by the stars. Centuries of work have finally pulled down the veil which had hidden from him the system of the world. Man is now seen to reside on a small planet, almost unseen in the vast extent of the solar system which, itself, is only a point in the immensity of space. The sublime consequences of this discovery are sufficient to give solace, despite the small space assigned to him in the universe, *by showing him his own nobility in having measured the sky from such a tiny base.*

Let us carefully conserve and continue to augment the body of this profound knowledge—the delight of thinking beings. It has rendered important service to navigation and geography, but its greatest benefit is to have dissipated the fears caused by celestial phenomena and to have destroyed the errors born out of ignorance of our true relationship with nature, *errors and fears which would promptly rise again if the flame of science were to be extinguished*—errors that are all the more distressing because our social order rests solely on these relationships. TRUTH, JUSTICE: These are her immutable laws. How far removed they are from our own dangerous maxim that it is sometimes necessary to mislead, deceive and subjugate man to ensure his welfare. Cruel experience inevitably proves that these sacred laws may not be violated with impunity.

Thus wrote one of the finest mathematicians in history. In preparing the fourth edition, published in 1813, Laplace decided to delete the final

---

* The two sentences thus marked were retained only in the first three editions.

remarks about truth and justice; he closed the book with the phrase "if the flame of science were to be extinguished." The surviving portion of that paragraph leaves no doubt that Laplace believed science to hold the key to man's salvation.

He died in a quiet country village at the age of seventy-eight, having preserved himself from the infirmities of old age by "strict abstemiousness," in the words of a nineteenth-century biographer. As he lay ill, facing death, he is reported to have said: "What we know is but little, what we do not know is immense."

14 THE SECOND HERSCHEL

Mathematics was John Herschel's major interest when he graduated from college, but he decided to take up law as a profession. His father wished him to join the clergy, because "Such a path must surely lead to happiness, or else it would never be so wide and so beaten"; but John objected: the church was based on self-deception and he would have none of it, he said. His father replied, "The miserable tendency of such a sentiment, the injustice and the arrogance it expresses, are beyond my conception."

Of the path of law William said: "It is crooked, tortuous and precarious. It is also beaten, but how many have miserably failed to acquire an *Honest* livelihood? You are not in that path, and it is almost too late to enter it; your studies have been of a superior kind." With a nice twist, he then said that lawyers must be deceptive at least half of the time, because "Whenever a lawsuit is decided the Lawyers on one side are always proved to be in the wrong, which must arise from ignorance, self-deception, or from a worse principle."

There was no suggestion of moral duty in William's arguments. He thought that a clergyman, "without the least derangement of his ostensible means of livelihood, has time for the attainment of the more elegant branches of literature, for poetry, for music, for drawing, for natural history, for short pleasant excursions of traveling"; he went on with a list obviously calculated to appeal to the tastes of his son.

In their next exchange of letters, William said the matter could not be settled until winter, when the two might converse at leisure. He added: "I can gather nothing from the contents of your letter but what amounts to, 'I do not like the Church; I prefer the Law,' and for this reason I cannot but wish to hear everything you have to say upon the subject. In hopes that you are perfectly satisfied of the affection of your father I shall only add that I am, very sincerely yours, Wm. Herschel." John prevailed, and in January 1814 he went off to London to read for the bar.

Eighteen months of that were enough. He then applied for the Chair of Chemistry at Cambridge, but lost the nomination. He took a job tutoring at St. John's, and in July 1816 obtained the degree of Master of Arts and was elected a faculty member at that college. He found teaching unbearable ("I am grown fat, full and stupid. Pupillizing has done this—and I have not made one of my cubs understand what I would have them drive at"), and he vowed not to take a position which would require him to teach again.

John visited his father that summer—at a time when his father was finishing what they both must have suspected to be his last important paper. William was seventy-eight, and both he and Caroline had nearly ceased observing. Yet, there lay before them the enormous task of completing the catalogues of nebulae and of double stars, and more measurements were needed to fill in the gaps. On October 10, 1816, John wrote one of his mathematical friends, Charles Babbage, that he had agreed to take up astronomy. He left no doubt how depressed he was:

I shall go to Cambridge on Monday where I mean to stay but just time enough to pay my bills, pack up my books and bid a long—perhaps last farewell to the University. . . . I always used to abuse Cambridge as you well know with very little mercy or measure, but, upon my soul, now I am about to leave it, my heart dies within me. I am going, under my father's directions, to take up the series of his observations where he has left them (for he has now pretty well given over regularly observing) and continuing his scrutiny of the heavens with powerful telescopes.

Six years later William Herschel died. Of the eulogies laid over his grave, there is one that seems particularly apt today. It was read by the French mathematician, Jean Fourier.

The influence of great men is prolonged into the future and the full fruition of their labours cannot be fully appreciated at their death. The picture of the heavens traced by William Herschel will be compared, by future generations, with more recent observations; some portions of the spectacle of the heavens may have changed, but in those distant ages the memory of Herschel will still endure. His name, confided to a grateful science, will for ever be safe from oblivion.

A month later Caroline left for Hanover, where she lived twenty-six more years—to a lively ninety-eight. She was in constant correspondence with her nephew, who visited her several times; and she garnered many awards, among them the Gold Medal of the King of Prussia.

John inherited a considerable fortune from his father—most of the money had been derived from the sale of telescopes—and for the next twenty years he remained an independent pursuer of optics, chemistry, and astronomy. In 1833, he packed up his family, then comprising a wife and three children, and his father's twenty-foot telescope, and set sail for the Cape of Good Hope, where he harvested the Southern nebulae and stars for four years. The telescope was erected in the country near a comfortable house and the family lived idyllically. Gunther Buttmann says:

Scarcely any branch of science escaped Herschel's attention during the four years at Feldhausen, though his interest was often sporadic and sometimes amateurish. Physical and chemical experiments are mentioned in his diary, his laboratory being a shed near the house which also contained the polishing equipment for his telescope mirrors. Stimulated by the delightful and interesting vegetation of the en-

The Clouds of Magellan, in the southern sky. Discovered by Magellan, these patches of light are detached from the Milky Way and are readily visible on a moonless night. The Large Cloud (*left*) contains many lesser nebulae, variable stars, and clusters of stars. The Small Cloud (*right*), although simpler in appearance, has nearly as wide a variety of constituents. Because these clouds are invisible from the Northern Hemisphere, they had been largely ignored until John Herschel pointed to their unique character. They are now regarded as irregular galaxies, the closest neighbors of the Milky Way but well beyond its borders.

vironment of Feldhausen, Herschel discovered a predilection for botany. . . . He brought back to England an impressive collection of
bulbs, most of which he presented to the Royal Horticultural Society
for further cultivation.

(Charles Darwin, voyaging on the *Beagle,* visited Feldhausen and was
struck by Herschel's modesty and apparent discomfort in society.)

The stay in Africa was a rich success astronomically; it was the
culmination of John's career as an observing astronomer. In a large
volume entitled *Results of Astronomical Observations Made During
the Years 1834, 5, 6, 7, 8, at the Cape of Good Hope,* he published
catalogues of Southern nebulae and double stars, reported observations
of sunspots and comets, and discussed the nature of the nebulae. His
remarks on nebulae are important because no one was better prepared
than he to discuss the problem, and they are interesting because they
reveal an important gap in his understanding of light and its origin. He
insisted that a gas cannot by itself emit light; only solid particles can
glow. He said, "Air, however intensely heated (if perfectly free from
dust), gives out no light. . . . The flame of mixed oxygen and hydrogen can hardly be doubted to owe what little light it possesses to inter-
mixed impurities." But in this he was wrong, as was proved twenty
years later in laboratory experiments with intensely heated vapors. We
know now that any substance, pure or impure, will give off light if it is
heated sufficiently: A tenuous gas will emit particular colors; a thick gas
such as the sun will emit *all* colors regardless of its composition. Even
the air will give off light if it is adequately excited, as the Northern
Lights attest.

John Herschel conceived the universe to be filled with tenuous
matter that could not glow but which carried a mixture of solids and
liquids that could glow. In an unresolved nebula the solid particles
might be very finely divided—and here he mentioned the luminous
appearance of the sea produced by tiny organisms—or it might consist
of very faint stars. There was no way of distinguishing, nor was there
any sharp line between the various degrees of subdivision.

Aside from the catalogues, the most significant result in this publication was his discussion of the Clouds of Magellan, two nebulosities
detached from the Milky Way in the Southern Hemisphere. Herschel's
naked-eye drawings show the smaller cloud as a roughly round patch
and the larger cloud as a faintly spiraling, elongated patch; they were

"pretty conspicuous" on a dark night, but the moon easily effaced them, he recorded. Within the clouds he found all manner of astronomical objects: single stars, clusters, and nebulae—an important result because all of these specimens must be at nearly the same distance from the earth and therefore their distinctive appearances must be due to differences in structure. He urged astronomers to follow him into the study of these clouds. They remained unstudied until well into the twentieth century due to lack of large telescopes in the Southern Hemisphere, but they are now recognized as systems outside our Milky Way and they have provided a valuable opportunity to compare different types of stars and clusters.

John Herschel quit observing when he returned to England. He was then forty-six, and he vowed that the publication of his results would mark the end of research for him. In fact, he never did erect the twenty-footer in England, and the forty-footer was ceremoniously dismantled and put to rest in 1840 with William Herschel's tools sealed inside the enormous tube.

During the decade following his return from the Cape, John alternated between the task of preparing his observations for publication and experimenting with the new technique he had dubbed "photography." When he heard that the tedious daguerreotype process had been supplanted by a process involving silver salts, his superb understanding of chemistry and optics put him in the situation of Galileo and the telescope: armed with the knowledge that it had been done, he did it for himself—he reinvented photography and made important improvements. The first photographic "negative" (his word) on glass portrayed his father's dismantled telescope. The scaffolding of the telescope made a fine test for the optics of his camera, and he no doubt chose it for this reason, but the picture nicely epitomizes the life of John Herschel: completion and preservation of his father's work, juxtaposed with inventions of his own.

In 1847 he finished the publication of his researches at the Cape and turned to the popularization of astronomy. He was delightfully successful, and his text, commonly known as "Herschel's Astronomy," went through twelve editions between 1849 and 1873 (if this is not a record, it is at least a sign of his gift for explanation). John's writing was facile and occasionally verbose; he was rarely colorful but he wrote with warmth and evident respect for the men whose work he reported, and the successive editions give a unique record of astronomy during that period.

With a collection of essays published in 1857, there is a short set of translations from Schiller and some original poems, so we cannot doubt that a romantic soul lay at the roots of his unruly hair. One poem is autobiographical; it is an apology:

> To thee, fair Science, long and early loved,
> Hath been of old my open homage paid;
> Nor false, nor recreant have I ever proved,
> Nor grudged the gift upon thy alter laid.
> And if from thy clear path my foot have strayed,
> Truant awhile,—'twas but to turn, with warm
> And cheerful haste; while thou didst not upbraid,
> Nor change thy guise, nor veil thy beauteous form,
> But welcomedst back my heart with every wonted charm.
>
> High truths, and prospect clear, and ample store
> Of lofty thoughts are thine! Yet love I well
> That loftier far, but more mysterious lore,
> More dark of import, and yet not less real,
> Which Poetry reveals. . . .

This poem is undoubtedly an allusion to his years in political office as Master of the Mint. At the age of fifty-seven, Herschel altered the pattern of a lifetime; he dove into public office, emerging four years later after a nervous breakdown. Buttmann suggests that Herschel's motivation may have been financial, or it may have arisen from a

> conviction that it was his moral duty to use his energy and capabilities to promote the public good of his country as well as his own inclinations of personal interests, even though the public service would mean giving up his scientific activities. The modest view that he always took of his scientific achievements led him to overlook the basic fallacy of this belief. He failed to realize that he could render far greater service to his country as a scientist than he could in public office.

Buttmann admits that "there will probably always be a certain degree of mystery as to Herschel's motives."

The poem tells us little, but it implies that Herschel thought the decision needed explaining. Perhaps we may see that poem as an admission of the sense of conflict between the life of science and of art within its author.

After retiring from the Mint in 1855, Herschel remained in seclusion until he died in 1871. His last years were peaceful and seemingly happy; he settled into the tedious labor of preparing the final catalogues of nebulae and double stars commenced by his father.

Julia Margaret Cameron, the famous English photographer, was a frequent guest of the Herschels' during this period, and her portrait of John (plate, page 163) is surely one of the finest products of early photography—how wild he looks by contrast with his father at the same age (plate, page 127).

John Herschel's assessment of himself, written to a friend in 1826 when he was thirty-four, professes a sense of amateurism. He said that it was partly out of a sense of vanity that he would prefer his contribution to knowledge to be regarded as that of an amateur rather than a professional scientist. He wrote: "possibly too [it is] a kind of obscure consciousness that I am not destined . . . to make giant inroads into great branches of human knowledge—but rather to loiter on the shores of the oceans of sciences and pick up such shells and pebbles as take my fancy for the pleasure of arranging them and seeing them look pretty."

This is a paraphrase of Isaac Newton* and I think John Herschel's comment reveals a profound difference from his father. John Herschel undoubtedly knew the world expected great things of him and he saw this as a burden on his life—a burden he magnificently carried in his astronomical work, but which he would have preferred to set down. (One night, obviously in a depression, he said he would break the mirrors and melt the telescopes.) If he professed amateurism and if he dashed from mineralogy to botany to chemistry, etc., perhaps he did this so the world could not catch up with him and pin a label on him.

John Herschel's death was mourned widely and deeply. In Buttmann's words:

He was mourned not only because of the loss of a man whose life had been so richly endowed with both intellectual and human values, but

---

* "I do not know what I may appear to the world; but to myself I seem to have been only like a boy playing on the seashore, and diverting myself in now and then finding a smoother pebble or prettier shell than ordinary, whilst the great ocean of truth lay all undiscovered before me."

John Herschel (1792–1871), only son of William. He spent four years in South Africa, extending his father's survey of the sky to the Southern Hemisphere. He then turned to popular writing and, briefly, to public administration. John Herschel made substantial contributions to the new process called "photography," a name which he invented.

also for the passing of an era that ended with his death. It had been an era of intellectual universality. . . . Herschel was by no means the only man of his time to embody the ideal of universal learning—there were many others both in England and elsewhere—but he was the one who embodied it most successfully.

It was also said of him that "He touched nothing that he did not adorn."

15 LORD ROSSE
FINDS SPIRAL NEBULAE

Born in 1800, William Parsons entered the English Parliament twenty-one years later, but the young nobleman had a special taste for mechanics and the construction of complex machinery, and he soon turned to astronomy. Sir Robert S. Ball, an astronomer who knew him in later life, says in a biographical note,

> I remember on one occasion hearing [him] explain how it was that he came to devote his attention to astronomy. It appears that when he found himself in the possession of leisure and of means, he deliberately cast around to think how that means and that leisure could be most usefully employed. . . . He came to the conclusion that the building of great telescopes was an art which had received no substantial advance since the great days of William Herschel . . . Thus it was he

decided that the construction of great telescopes should become the business of his life.

In 1827, Parsons started a series of experiments in the casting of large metal mirrors, taking up where Herschel had left off because he was convinced that large transparent lenses—the alternative—could not compete with mirrors. History proved him right. The largest successful lenses were made a half-century after Parsons; they were but forty inches in diameter, and they collected only one-fourth the light of Parsons' telescope. Lenses are more difficult to manufacture than mirrors because the light must pass through them rather than reflecting off the surface. A surface may be polished with relative ease, but the internal strains in glass distort the passing light, and they are particularly treacherous in large lenses.

Parsons made his mirrors of speculum, an alloy of tin and copper used by Herschel. It had a reflectivity almost as high as that of silver—two-thirds of the incident light was reflected, one-third was absorbed, if the surface was polished. The casting of these mirrors was a challenge; if the alloy mixture was adjusted to give a higher reflectivity, the mirror became brittle and was quite likely to break during manufacture—so the choice was a gamble. Parsons' description of the task facing a hopeful mirror maker was rather bleak:

> He would of course first proceed to cast the metal. As earthen vessels would not be sufficiently capacious, he would employ either iron ones or a reverberating furnace. If he tried iron vessels, before a large quantity of speculum metal, for instance three or four hundred weight, was raised to a proper heat for casting, he would find that the metal had imbibed some of the iron, and was injured; or perhaps, if he was less fortunate, and the fire had been a little mismanaged, that the speculum metal had promoted the fusion of the iron, and so passed out through the crucible. The reverberatory furnace would then be resorted to. Much difficulty would occur in combating the continual change of the quality of the metal from the exposure of so large a surface to the action of the flame. However, the metal once cast, the next process would be to anneal it. He would then find that the speculum would fly to pieces before it was cool, unless the alloy made use of was less bright, less white, and in every respect inferior to the best speculum metal.

A colleague wrote: "This compound is brittle almost beyond belief; a slight blow, or even the application of partial warmth, will shiver a large mass of it; though harder than steel, its surface is broken up with the utmost facility, and it has a most energetic tendency to crystallize."

Parsons discovered that he could overcome the worst of the difficulties by constructing a laminated mirror pieced together from thin, pie-shaped sections of bright speculum laid over a base of crude, strong alloy. In his own laboratory, with the aid of men he had trained, Parsons constructed a thirty-six-inch mirror and mounted it in a tube twenty-six feet long. This telescope collected about four times as much light as Herschel's work horse (the twenty-footer with a nineteen-inch mirror), and its surface was considerably brighter. Whether the performance of this telescope exceeded that of Herschel's forty-footer, Parsons was never sure. (There were some who said Herschel's forty-footer was a failure, and the persistence with which Herschel's friends protested this allegation suggests that indeed it may have been true.)

Parsons left Parliament in 1834, and in 1840 the first description of the performance of the thirty-six-inch telescope was published. A friend, the Reverend Thomas Robinson, presented this report to the Royal Irish Academy with the remark, "It is scarcely possible to preserve the necessary sobriety of language, in speaking of the moon's appearance with this instrument."

Parsons became the Third Earl of Rosse upon his father's death in 1841, and his success with the thirty-six-inch telescope encouraged him to start on a telescope with a seventy-two-inch mirror and of fifty-foot focal length. He cast the speculum in April 1842, and began the gigantic task of grinding a half-inch depression in the six-foot disk of metal. Within a year this task had been finished and the foundations for the mounting had been laid at Birr Castle in Northern Ireland. Another year saw the telescope mounted, and the polishing of the mirror began in the summer of 1844. By February of 1845 "the work was sufficiently advanced to permit use of the instrument without personal danger." The upper end of the fifty-foot tube was hooked to a chain cable running through a fixed pulley to a windlass on the ground, which was worked by two men. The entire structure was suspended between two concrete piers, one standing to the east and the other to the west of the telescope, so the telescope was confined to the sky along a narrow band from the southern sky overhead to the northern sky.

Sir Robert Ball described the impression the telescope made on visitors:

> On an extensive lawn, sweeping down from the moat towards the lake, stand two noble masonery walls. They are turreted and clad with ivy, and considerably loftier than any ordinary house. As the visitor approaches, he will see between those walls what may at first sight appear to him to be the funnel of a steamer lying down horizontally. On closer approach he will find that it is an immense wooden tube, sixty feet long, and upwards of six feet in diameter. It is in fact large enough to admit of a tall man entering into it. . . . This is indeed the most gigantic instrument which has ever been constructed for the purpose of exploring the heavens.

But the instrument was striking not only for its sheer size, but because it was totally different from other telescopes of its day. "The astronomer at Parsontown has . . . to avail himself of the ingenious system of staircases and galleries, by which he is enabled to obtain access to the mouth of the great tube."

Even before the instrument was completed, the men set to work looking for the Great Nebula in Orion. But the weather was foul, and they saw nothing for a month; finally, they removed the mirror from the telescope for its final polishing.

With the advent of better weather, Rosse and Robinson started attacking one by one the nebulae on John Herschel's lists. They soon found that they could resolve many of them into individual stars, and this success suggested to them that all nebulae might be composed of stars. They wrote: "There must always be a very great number of clusters, which from mere distance will be irresolvable in any instrument; and if it prove to be the case that *all* the brighter nebulae yield to this telescope, it appears unphilosophical not to make universal Sir J. Herschel's proposition, that 'a nebulae, at least in the generality of cases, is nothing more than a cluster of discrete stars.' "

One object was of particular interest. The fifty-first nebula of the catalogue of Messier and Méchain had been described as follows: "Very faint nebula without stars. . . . M. Messier discovered this nebula on the 13th of October 1773, while observing the comet which appeared in that year. Seen only with difficulty in a 3½ foot telescope. Reported on the chart of the comet of 1773–74. It is double, each having a brilliant center. . . . The two atmospheres touching; one fainter than the other. Reviewed several times."

In 1833, John Herschel published a diagram of this object, showing it as a ring with a central condensation: The ring was double in its southwestern part. Herschel noted that "supposing it to consist of stars," the ring would look exactly like the Milky Way to an observer inside. He then posed the question: "Can it be, then, that we have here a brother-system bearing a real physical resemblance and strong analogy of structure to our own?"

This was one of the first suggestions that the Milky Way is a ring of stars with the sun located inside. By the end of the nineteenth century most astronomers had accepted the idea, because it nicely explained the fact that the sky reveals a scattering of bright stars in the foreground and clouds of very faint stars hanging in the background.

With the seventy-two-inch telescope, Rosse verified Herschel's impression of "M51"—as the object was then known because it was fifty-first in the list of Messier and Méchain. He said, "Were the center absent, we should have a ring nebula; and were the line of vision near the plane of this ring it would become one of those rays with a bright nucleus and parallel band or satellite nebulae which occur so frequently in the catalogue [of J. Herschel]."

Shortly after the first of Rosse's sorties into the sky, a devastating potato famine struck Ireland; blight followed by typhus swept the country, and for two years Lord Rosse abandoned astronomy because, according to Robinson, he was "not a person to seek knowledge or enjoyment in the heavens, when he ought to be employed on earth; and he devoted all his energy to relieve the present misery and provide for the future." He returned to the telescope three years later, in 1848, when the "days of evil" had passed.

Re-examining M51, he realized that this and several others showed a spiral pattern. With great care, he measured and drew these spirals in the hope of detecting the swirling motions with the passage of the years. He did not succeed in detecting motions.

His drawing of M51 shows a distinct spiral; it is far more complex

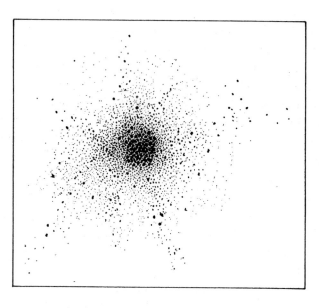

Drawings by the Earl of Rosse. A globular cluster is resolved into stars (*above*), and the filamentary structure of the Crab Nebula is evident. It was on the basis of this drawing that the Crab obtained its name. Compare this rendition with the photograph in red light shown in the plate on page 171. The human eye is less sensitive than the plate to the radiation of the filaments.

The Crab Nebula, M1. The impression of outward motion has been verified by measurement of the filaments. This nebula is the residue of a stellar
explosion that occurred in the year A.D. 1054. The arrow points to a
"pulsar," which is evidently the remnant of the original star. The pulsar
emits radiation of all types—from x-rays to radio waves—and all of its
signals pulsate more than a hundred times per second.   (*Lick Observatory,
University of California*)

The Earl of Rosse's drawing of the spiral nebula M51. With his 72-inch telescope Lord Rosse discovered that the pattern was a spiral, and he subsequently found many others. Compare with the modern photograph on page 173.

than the simple ring seen earlier. In commenting on observations by others and himself, Rosse noted the progressive ·elaboration of the visible pattern and said:

> We thus observe, that with each successive increase of optical power, the structure has become more complicated and more unlike anything which we could picture to ourselves as the result of any form of dynamical law. . . . The connection of the companion with the greater nebula . . . adds . . . to the difficulty of forming any conceivable hypothesis. That such a system should exist, without internal movement, seems to be in the highest degree improbable.

The announcement of spirals in the nebulae caused a great stir among astronomers because they hoped to find a clue to the nature of these objects—their internal forces and motions—from the patterns.

The conclusion that the spirals were rotating seemed inescapable, but Rosse refrained from speculation, insisting that the data were too sparse. He evidently believed the spiral nebulae to be composed of clouds of stars. He wrote:

On the dynamical condition of such systems it would at present be idle to speculate; it must evidently be much more complicated than that of

M51, a spiral nebula. This photograph, taken with the 120-inch telescope of the Lick Observatory of the University of California, should be compared with Lord Rosse's drawing, shown in the plate, page 172. This is considered to be a double nebula, and the outer arm of the larger nebula has evidently been distorted by the companion.

The spiral nebula M81. The striking symmetry of this two-armed spiral is quite typical. Fine, dark lanes of dust may be traced into the central regions, and numerous bright stars and clusters may be seen in the outer arms. (*200-inch photograph, Hale Observatories*)

the ordinary globular clusters, which themselves are intricate enough. Their resemblance to bodies floating on a whirlpool is, of course, likely to set imagination at work, though the conditions of such a state are impossible there. A still more tempting hypothesis might rise from considering orbital motion in a resisting medium; but all such guesses are but blind.

Later he wrote, "When certain phenomena can only be seen with great difficulty, the eye may imperceptibly be in some degree influenced by the mind; therefore a preconceived theory may mislead, and speculations are not without danger." The later history of spiral nebulae showed only too well the influence of the mind on the eye; Rosse had found several spirals where spirals were no longer credited. On the other side of the coin is the fact that many astronomers found spiral structure with their own smaller telescopes *after* Lord Rosse had pointed them out.

In 1843, Lord Rosse was elected president of the British Association for the Advancement of Science, an honor he accepted with the

> awe which it is impossible not to feel in the presence of men the most distinguished in the varied departments of human knowledge. . . . This very embarrassing position is not of my own seeking. To have aspired to the high honour of presiding at one of your meetings would have been an act of presumptuous vanity, which I never did, which I never could have contemplated. A communication from Manchester, announcing that the Association had actually made their selection, was the first intimation which reached me that my name had even been thought of.

Today, it would be astonishing for a president to be unaware that he was under consideration, and it is difficult to imagine how it could have occurred in Rosse's time. His confession, and the fact that Robinson so often spoke for him, give the impression that Rosse was preoccupied with other matters. His "awe" and his sense of being put into a "very embarrassing position" suggest that he felt like an outlander who had been drawn into a sacred circle of science. He rose to the occasion gracefully with a simple address of the "triumphant" role of the British Association in fostering science.

It is hard to avoid thinking that Rosse was describing himself when he said, "The man of the world who, busied in the changing scenes of life . . . cannot fail to look with surprise, and I may add, with gratification, at a meeting so large (and in this country too), from which politics are altogether excluded."

Rosse's astronomical observations spanned twenty years, although he entrusted the telescope more and more to his associates as the years progressed. Sir Robert Ball wrote: "I think that those who knew Lord Rosse well, will agree that it was more the mechanical processes incidental to the making of the telescope which engaged his interest than the actual observations with the telescope when it was completed. Indeed one who was well acquainted with him believed Lord Rosse's special interest in the great telescope ceased when the last nail had been driven into it."

The drawings of nebulae made by Rosse and his collaborators became outdated by photography shortly after his death, but Rosse's telescope had revealed a new species of astronomical objects: spiral nebulae. He had scrutinized the forms of the nebulae, measuring the location of knots, the ends of wisps, and the shapes of curving filaments of light. He demonstrated once again that close examination of the sky reveals delightful complications. In this, and in his restraint from theorizing, Rosse's scientific career revealed how far modern astronomy had come from that of the ancient Greeks.

The great telescope and the optical shops about it attracted visitors from all over the world. Sir Robert Ball wrote: "His home at Parsontown became one of the most remarkable scientific centres in Great Britain; thither assembled from time to time all the leading men of science in the country, as well as many illustrious foreigners."

His later years, however, were spent in relative seclusion broken by trips to London during "the season" and by cruises on his yacht. He occupied himself with studies of political and social questions, and died at sixty-seven.

PART III
RESOLUTION OF THE ENIGMA

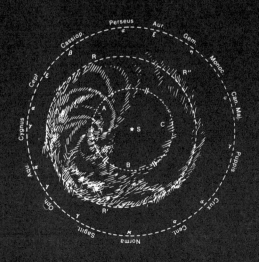

16 BIRTH OF THE NEW ASTRONOMY

In the early nineteenth century one philosopher declared that man would never know the chemistry of the stars. Yet, by 1859 dozens of chemical elements had been identified in the sun and stars; within another thirty years, a previously unknown element had been detected on the sun and tracked down to earth in a laboratory. It began with Isaac Newton.

Newton discovered that sunlight could be separated into component colors by transmission through a glass prism. Light of each color had a different "refrangibility," so it was bent through a different angle by the prism surfaces. When the light was focused again on a screen it produced a colored ribbon, arranged from blue through green and yellow to red: the "spectrum" of sunlight. Newton did not think of this as a means of studying the sun; his attention was focused on the behavior of the light, and he soon realized that he had uncovered the main defect of telescopes of his day. They all used a lens to focus starlight, and they separated the red light from the blue; for this reason the images were unclear. Newton devised a telescope of mirrors because he knew that all colors *reflect* from the surface of a mirror at the same angle.

Opticians later discovered that by combining lenses made of different glasses they could greatly reduce the unwanted separation of colors. Joseph von Fraunhofer, a German optician, became dissatisfied with the quality of his lenses and sought a way to make more precise measures of the refrangibility of each color. Light from a yellow flame was useful, because it was monochromatic; that is, it consisted of a single color. But Fraunhofer needed measures in all colors, so in 1814 he spread out the light of the sun with a prism, hoping to find other narrow bands.

He found hundreds of narrow dark regions in the spectrum of the sun. A few years later he examined the light of the stars and planets to verify that his lenses would work with their light as well as that of the sun, and he found the light of the planets to have many features of the sun's spectrum.

The stars were another story: each spectrum had its own set of lines. He found, however, that the various colors constituting starlight were refracted in the same manner as the colors of sunlight, and having confirmed that his lenses would work he turned to other matters, with the remark: "In all my experiments I could, owing to lack of time, pay attention only to those matters which appeared to have a direct bearing on practical optics."

At about this time, John Herschel had found another use for the prism. He discovered that each chemical element displayed a characteristic set of colors when placed in a flame. Examining the light with a device known as a spectroscope—a combination of prisms and a small telescope—Herschel found that he could identify slight traces of, for example, the salts by noting the colors.

Herschel was among the first to offer a partial explanation of the interesting coincidence noted by Fraunhofer: The refrangibility of the yellow light emitted by a sodium flame equaled that of the yellow light in the sun's spectrum near a dark pair of lines Fraunhofer had labeled "D" in his map. Herschel supposed that an atom is like an organ pipe which can resonate at a particular frequency and emit a musical note. If a tuning fork of the proper pitch is brought near an organ pipe, some of the fork's vibration energy will be absorbed by the organ pipe, and the pipe will sing in response. In the same way, light passing through matter, such as colored glass, excites the atoms into oscillation, and is weakened.

By the middle of the nineteenth century it was possible to go a step further in this description and say that heated matter emits precisely the same colors that it absorbs, just as an organ pipe can emit the note that it absorbs. Further, each chemical element has a characteristic set of frequencies to which it can respond and which it can emit. This is the underlying principle of spectrum analysis. Although further experiments showed that the absorption and emission of light depend also on the temperature of the gas, they confirmed the one-to-one relationship between an element and its spectrum.

Gustav Kirchhoff, a physicist of Heidelberg, and the chemist Robert Bunsen charted the dark lines of the solar spectrum and compared them with lines emitted by various chemical elements. In 1859, they identified over two dozen atoms in the atmosphere of the sun, thus opening a new era in astronomy.

William Huggins, a wealthy English amateur, had purchased an eight-inch lens and built an observatory at his house in 1859. For a while he mapped sunspots and made drawings of the moon and of Jupiter, but soon he tired of the "routine character of ordinary astronomical work," and in a "vague way" sought work in a new direction. Then word came of Kirchhoff's great discoveries and the analysis of sunlight.

Huggins later described his reaction: "Here at last presented itself the very order of work for which in an indefinite way I was looking— namely, to extend his novel methods of research upon the sun to the other heavenly bodies." Huggins was prepared for this by previous training in chemistry, and he found a friend willing to collaborate with him.

Together they built a spectroscope, attached it to the telescope, and equipped it with a small prism with which they could compare the star's spectrum to the spectrum of a terrestrial source. Then they set to work. Huggins wrote: "The time was, indeed, one of strained expectation and of scientific exaltation for the astronomer, almost without parallel; for

Sir William Huggins (1824–1910). A pioneer in the analysis of starlight, Huggins discovered that planetary nebulae show the spectrum of a gas while spiral nebulae show a spectrum resembling that of a star. He is shown here in his private observatory. At the lower end of the telescope is his spectroscope for dispersing light into its component colors. Prisms in the cylindrical container at the lower end produce the dispersion and send the light to the eyepiece in Huggins's hand. (*Reproduced from* Publications of Sir William Huggins' Observatory, *Vol. II* [*1909*])

nearly every observation revealed a new fact, and nearly every night's work was red-lettered by some discovery."

For over a year Huggins concentrated on the light of the brightest stars. Then he turned to the nebulae, and the first he examined was a bright planetary nebula in the constellation Draco. He was not prepared for what he found: "I looked into the spectroscope. No spectrum such as I expected! A single bright line only!" The light of the nebula, unlike that of the sun, was composed of a single narrow band of color. "Unlike any other that I had as yet subjected to prismatic examination," it could not be stretched out to form a complete spectrum of colors.

Closer examination revealed several other faint lines of light, and Huggins realized that he had discovered a way to settle the enigma of the nebulae. If they were stellar they would show a complete range of colors like the sun; if they were gaseous they would show only discrete lines like a flame. "A few days later I turned the telescope to the Great Nebula in Andromeda. Its light was distributed throughout." He concluded that it must be composed of myriad faint stars, and he examined sixty other nebulae and clusters, finding about one-third to display a bright-line spectrum like that of the planetary nebula. Some nebulae were gaseous and others were stellar; the spectroscope could distinguish what the ordinary telescope had not.

In 1866, Huggins built a more powerful spectroscope and achieved an entirely new type of measurement: he determined the velocity with which a star moves toward or away from the earth, known as its "radial velocity." Thus, he added a new dimension to the study of stellar motions.

Light, like sound, travels at a measurable speed, and it has some of the properties of a stream of waves. One such property is the change of observed frequency—the pitch or the color—when the source of the waves is in motion with respect to the observer. (The most familiar terrestrial example of this shift, named for Christiaan Doppler, is the sudden fall in the pitch of a passing train whistle.) If a star is in motion away from the observer, all features of its spectrum are shifted toward the red. By an extraordinarily delicate comparison of the light from terrestrial sources and the light from stars, Huggins was able to demonstrate that stars move toward or away from the earth at speeds of ten or twenty miles per second.

O5-B0 STANDARDS

The System is that of H.H. Plaskett (O5-9)

Pubs. D.A.O. 1,365,1922

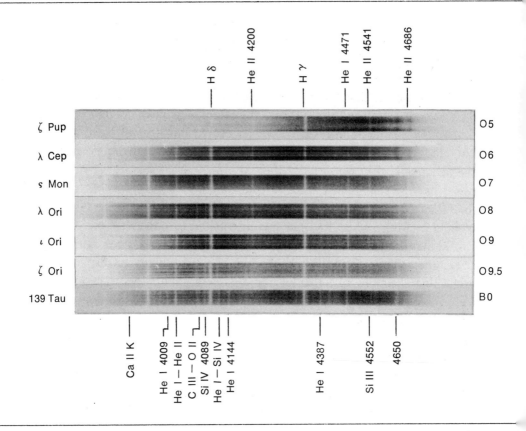

The principal criterion of type is the ratio He I 4471: He II 4541 At O9.5 λ 4200 is still visible and the ratio He I 4387: λ 4200 is used. At B0 the He I spectrum is stronger in general, while the line Si IV 4089 is stronger than Si III 4552.

Eastman process

Photographs of stellar spectra. The light of each star is dispersed according to wavelength. This is a negative photograph, so the light lines actually represent missing light in the spectrum of each star. Note that the stars have been arranged in a sequence of gradual changes; more than 99 per cent of all stars can be fitted into such an array. The position of a star in the sequence reveals the star's temperature and gives a hint as to the star's size. This illustration is from an atlas of stellar spectra used by astronomers to classify stars; the classification criteria are listed for the convenience of the astronomer. (*Yerkes Observatory, University of Chicago*)

The nineteenth century was not entirely a century of wealthy amateurs, but there is no denying Huggins's feeling that he and others like him who searched for new techniques were singing "a very lovely song of one that hath a pleasant voice." Another such man was Henry Draper, born in 1837 to John William Draper, an eminent physician and chemist of New York City. Henry's mother was the daughter of the attending physician to the Emperor of Brazil and, like John Herschel, the young boy was stimulated to precocity in the company of his parents and their friends.

When he was twenty, Henry visited the Earl of Rosse's mammoth observatory and conceived the possibility of combining astronomy with photography. Returning home, he joined the staff of the Bellevue Hospital and used his spare time to grind and polish a speculum mirror with the techniques he had learned from Rosse. But the mirror split.

His father told John Herschel of the boy's frustration, and was advised to suggest that Henry build a mirror of glass and then coat it with silver. The method of grinding and polishing was not very different from that for a speculum mirror, but the glass was much lighter and less brittle and his efforts quickly brought success. Within a year he had built the first of his one hundred mirrors and had installed it in an observatory at his father's estate on the Hudson River.

Henry was a mechanical perfectionist, and he built six clocks to drive his telescope before he was satisfied. In 1864, the Smithsonian Institution published his monograph "On the Construction of a Silvered Glass Telescope of $15\frac{1}{2}$ Inches Aperture, and its Use in Celestial Photography." It became the standard reference for telescope makers.

Wet collodion plates with all their attendant annoyances were still in use for photography when he started, but in 1879 Draper visited Huggins in England and learned that dry plates had become sufficiently sensitive to be used in a telescope. He mounted these new plates in his telescope and succeeded in photographing the moon, the brighter nebulae, and even the spectra of several stars. His techniques were soon adopted throughout the world, and astronomers were at last able to obtain permanent and objective records of the sky.

Henry Draper died at the age of forty-five from pneumonia, which he contracted following an eclipse expedition to the Rocky Mountains with Thomas Edison, Mrs. Draper, and two scientific friends. His wife's memorial gift to the Harvard College Observatory initiated what became, and will probably remain for some time, the most comprehensive survey of the spectra of stars—encompassing almost half a million stars and providing a rich resource for astronomers of the twentieth century.

There is a strong similarity between starlight and the light emitted by a furnace, so we begin a discussion of starlight by imagining a furnace that is fully stoked and burning well; the coals and the interior walls are red. If the draft is increased and the fire becomes hotter, the walls and the coals will turn orange, or perhaps yellow. Nothing is visible inside; all is a uniform orange glow. Throw in some coal; it will appear dark for a few moments and then it will warm up and take on the orange color of its surroundings; it will disappear long before it has burned up.

A piece of blue glass held in front of the window absorbs the red light from the furnace, passing only the blue light. If the glass were placed inside the furnace, it would appear blue for a moment and then it too would vanish in an orange glow. As the glass heats up, it begins to radiate the red light; it radiates more and more and we see its color as a mixture of the blue that passes through and the red that it emits. The combined light becomes precisely the orange of the furnace, so the glass becomes invisible.

This behavior of matter in a furnace is not an intuitively obvious pattern; but it is a fact of nature. It is observed and it must be taken as the starting point for theories of heat radiation. From the fact that the glass vanishes, we infer that it radiates the colors that it absorbs. This is the law by which Kirchhoff described the behavior of atoms: they radiate the colors that they emit. He went a step further and said that the ratio of the absorptive and emissive power in each color is independent of the type of material.

From the fact that the glass, or whatever we put in the furnace, *can* come to equilibrium with the walls, we can infer something else. Equilibrium implies a static, timeless state; appearances are unaltered from one moment to the next. Every detail is precisely as it was a second or a year before; this timelessness has a profound implication and a fancy name: "the principle of detailed balancing in thermal equilibrium." The name arises from the fact that thermal equilibrium is achieved when each particle in the furnace radiates exactly as much as it absorbs in every color. This principle is one of the simplest and most powerful of the ideas underlying twentieth-century physics. From it, physicists were able to derive laws describing the light emitted by a star and the light trapped within a star. For these reasons we will detour slightly into the game of poker, and construct an analogy to the principle of detailed balancing.

Suppose Alphonse and Bob play poker with chips of three denominations: blue chips worth 10¢, yellow chips worth 5¢, and red chips worth 1¢. At the start of the game each player has equal money in chips, and the players are perfectly matched so that after playing for an hour they are still equally rich. (After each hand, we may expect one player to be richer than the other, but these fluctuations occur in both directions.) Now in order to make this game analogous to the state of thermal equilibrium within the furnace, when all distinctions vanish, we must suppose that it is impossible to identify the players merely by watching the chips. In other words, if the players wore masks we could not say, for example, that Alphonse is the one who gives more of the blue chips to Bob. The assumption that this will happen is equivalent to the assumption that thermal equilibrium can be achieved in the furnace.

What must occur, if the players are to be indistinguishable? In the long run, each player must give precisely the same number of *each type* of chip to the other player. This is the principle of detailed balancing in equilibrium.

These ideas were developed early in the nineteenth century, but physicists saw that they were incomplete. They did not suffice to predict, for example, the apparent color of the furnace for each temperature. (In the same way, our present description of the poker game does not define the number of blue, yellow, and red chips held by each player.)

Two further assumptions were added at the close of the nineteenth century, and these permitted a prediction. The first was a very general assumption describing the storage of heat in the walls of a furnace: Heat is stored in oscillations of the atoms; the greater the temperature, the greater the quantity of heat and the more rapid are the oscillations. In terms of the poker game, we would say that a "hotter" game is one in which the players have more money and play for higher stakes.

One of the central problems is to establish the manner in which the relative number of chips varies with the total amount of money, because this would be analogous to the change of the color of radiation with the temperature of the furnace. In order to effect this calculation a more complete description of the game is needed.

We introduce the assumption that every step of the poker game (the original distribution of chips, the play of cards, the choice of chips for payment of losses, etc.) is taken in a random fashion according to chance. This will guarantee that the players will be equally matched.

Physicists assumed that the heat stored in the walls of a furnace was apportioned to red, yellow, and blue bits of energy, but when they computed the actual amounts they found that the predicted color was not what they saw in the furnace.

The difficulty was resolved in 1900 when Max Planck, a German physicist, suggested that the analogy between a furnace and a poker game be taken more literally than it had been. He pointed out that the physicists had assumed heat to be stored in an infinite number of different bits—as though the banker had an infinite variety of chips rather than just three types. Planck proposed that the computation should stick to the rules of poker and recognize that heat energy comes in finite lumps.

This was the quantum hypothesis: Heat can be stored only in a finite number of ways within the walls or within a star. This viewpoint sufficed to make the calculation succeed.

By this time astronomers were measuring the light from stars. They found some stars to be bluer than others, and they noted that the general quality of starlight was not very different from that of furnace light. They wished to derive temperatures for the stars, so they assumed that the inside of the star was like the inside of a furnace and could also be described by Planck's calculation. They further assumed that the light escaping from a star could be matched by the light within a furnace, and they derived temperatures of 30,000°C to about 3000°C for the stars. The sun was found to be about 6000°C.

Serious discrepancies appeared as the measurements improved; the spectrum of starlight did not have the smooth distribution among colors that furnace light had. After another thirty years the discrepancies were successfully attributed to the fact that the surface of the star is exposed to the darkness of space rather than being enclosed within a furnace.

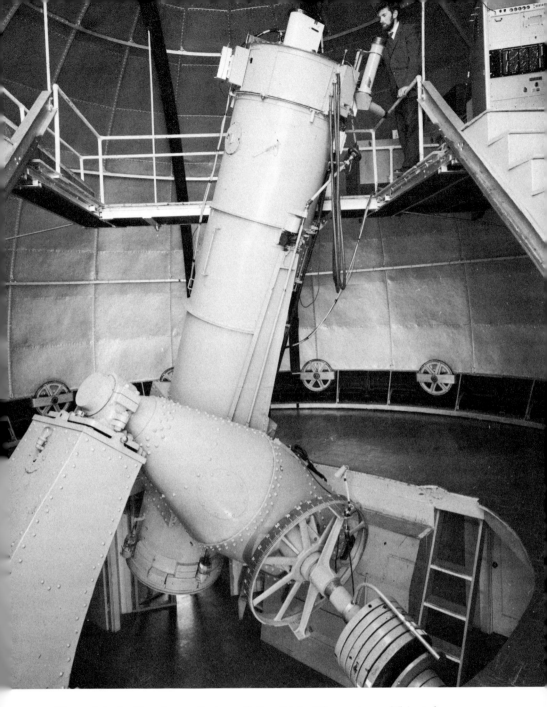

The 36-inch Crossley reflector of the Lick Observatory. This telescope made many of the fundamental photographic researches on the Milky Way during the last decade of the nineteenth century and the first several decades of the twentieth century. It is still in operation and is used for many projects not requiring a larger mirror. (*Lick Observatory, University of California*)

The calculations are now exceedingly complex, but they permit a nice correlation between the quality of a star's light and the temperature and gas pressure within its atmosphere.

Such were the concepts with which astronomers were equipped at the close of the nineteenth century.

17 ASTRONOMY ON
THE VERGE OF A NEW ERA

By the start of the twentieth century, much astronomy appeared to be in the hands of taxonomers. Stars were classed according to color, spectrum, brightness, and motion. They were counted; their distances were measured by triangulation, using the earth's orbit about the sun as a baseline; their true brightnesses were computed from their known distances; their dimensions were estimated from their total brightnesses.

Astronomical observatories published vast tomes containing lists of stars and their properties, but no scheme had been developed which was capable of encompassing the enormous variety of stars, clusters, nebulae, dark clouds, and rocks that had been found in the sky.

There was no organization of astronomers to recommend uniform notation or coordinated programs of research, so each observatory went about its work as it saw fit. Chaos was the result. By 1900, a single star could be described in more than a dozen different ways, depending on the observatory or the publication from which the information had been culled.

The age of the earth had been estimated in two different ways. Geologists examined the conformation of rock strata and the location of fossils—they concluded that the earth must be billions of years old. Astronomers estimated the age of the sun—*they* concluded that it could hardly be more than one hundred million years old, if that much. And if the sun were that young, they said, the earth must be younger.

The expected lifetime of a star can be estimated in the way we guess when a furnace will run out of fuel: the supply of fuel is compared with the rate of consumption. The rate at which a star consumes fuel is equal to the rate at which it pours light into space, because the energy of a star escapes as light. Estimates of consumption were made slightly uncertain by the fact that the ultra-violet and infra-red portions of starlight are absorbed by the earth's atmosphere, but the blockage could be estimated.

From the outset, the crucial uncertainty lay in estimating the fuel available to a star. By the end of the nineteenth century, physicists had described the behavior of large masses of gas, but the behavior of single atoms was a total mystery. Atoms were still considered to be inscrutable points of matter—as they had been by the ancient Greeks. Two sources of energy were known at the time. The chemical energy of interaction among atoms appeared to be adequate for comets but it was inadequate to supply the energy of a star. However, it appeared that sufficient gravitational energy could be released by the slow contraction of a star into itself—its own matter gradually compacting under the influence of gravity.

In the 1850's H. von Helmholtz and Lord Kelvin (William Thomson) developed a simple and compelling theory describing the contraction of a star based on the conception of a star as an isolated body of inert gas. No energy was fed from the outside; no significant amounts of energy were released by chemical processes, and nuclear processes were unknown. Heat and light came from the pressing together of the gas by the force of gravity.

Lord Kelvin assumed that stars were born as enormous, relatively cool balls of gas, which compacted and warmed until they reached the temperature of luminescence. Then they cooled and shrank toward death as faint red stars. All of this was speculation. There was little evidence, and what there was could be dismissed as circumstantial, but the theory had the virtue of neatness and it permitted calculations of the lifetime of a star. It was this calculation that brought the geologists and astronomers to loggerheads.

Most astronomers must have simply shrugged their shoulders, believing that the discrepancy would vanish when more facts had been unearthed, but not all astronomers were willing to accept this state of affairs. Some sought other ways to feed the stars. The most persuasive of these men was Sir Norman Lockyer, a contemporary and fellow-countryman of Lord Kelvin. Lockyer was an experimenter; he developed one of the first laboratories devoted to the correlation of starlight with the light from electrical sparks and high-temperature arcs and flames. He examined the light emitted by fragments of meteoritic material in his laboratory, and he concluded that the presence of metal—predominantly iron and nickel—in the spectra of meteorites *and* stars proved that stars were surrounded by clouds of meteoritic particles. He suggested that the heat of stars and nebulae, including spiral nebulae, could be derived from solid matter falling swiftly together under the attraction of gravity. According to his picture the heat of a star is derived from the motion of matter, in much the way that the brakes of a vehicle are heated when the forward motion of the vehicle is arrested, or the way a high-speed projectile may vaporize when it strikes a solid target.

His theory exploited one of the great scientific achievements of the nineteenth century: the kinetic theory of heat, which states that heat is a form of motion distributed randomly among the microscopic elements of the hot body; high temperature was attributed to rapid motion of the atoms. Just as a projectile heats when its forward motion is converted into chaotic oscillations by impact against a wall, so the outer layers of a star might be heated by the impacts of infalling meteorites. His physical argument is still accepted; in fact his theory was recently resurrected to explain the high temperature of the sun's outer atmosphere (unsuccessfully), and a similar idea has been developed for quasars; but at the time it appears to have been treated as though it were a gigantic freak. In his first exposition of the idea, Lockyer attempted to encompass every object in the sky. He explained comets, bright nebulae and dark;

he saw variable stars as swarms of meteorites; the sun was fed by meteorites; novae revealed collisions between two great clouds. There was nothing he did not explain.

Lockyer's theory was ignored by some and hotly debated by others, but very few were willing to elicit inspiration from his theory. Lockyer himself treated it as a working hypothesis—a tool for stringing together the empirical facts and for giving direction to his experiments. Theories were mere conveniences for him, and he preferred a bad one to none at all.

At the close of the nineteenth century a large number of astronomers were attempting to determine the life history of stars and to interpret the data of color and size as evidence of age and origin; others sought to determine the "construction of the heavens." Some used techniques similar to those of the Herschels, but the assumption that all stars were of roughly the same brightness had been discarded, so the problem had become more complex.

From the number of stars of each apparent brightness in the sky, astronomers wished to determine the number of stars of each true brightness and the number of stars at each distance from the earth. This double problem cannot be solved unambiguously without further information, because increased distance reduces the *apparent* brightness of a star as does a reduction in the star's *true* brightness.

Triangulation had been achieved, but it was enormously difficult and only one hundred stars had been reached this way by 1900. So astronomers turned to an indirect method.

The spectroscope permitted measuring the speed with which individual stars approach or recede from the earth in their flights through space. The reddening of light from a receding star and the converse alteration from an approaching star are known as the "Doppler effect." The shifts of color are minute, being totally undetectable to the eye, and they require microscopic measurement on photographic plates.

The data accumulated slowly, but the consequences were revolutionary. For the first time, astronomers could measure a quantity that was not influenced by the distance of the star. The trick was to combine this measurement with a quantity that *did* depend on distance, and in this way to determine the distance. It worked in the following manner.

. . .

Over a century earlier, William Herschel had suggested that the nearby stars appear to move as a group across the face of the sky because the sun moves through space among them. He determined the velocity of the sun, but his results were received with skepticism because the data were very sparse. By 1900 the data had become abundant, and the sun's motion had been well established.

However, the sun was not the only star to move with respect to its neighbors; all stars did. Each star was discovered to have its own "proper" motion, and the proper motions were distributed more or less at random—or so it seemed at first.

Using the assumption that the proper motions were random, and combining them with the speeds determined by Doppler shifts, astronomers determined the distances of groups of stars by taking the ratio of random speeds through space to random apparent motions across the face of the sky. This technique was not very different from the calculation that a duck hunter makes, perhaps unconsciously, when he estimates the distance of a duck by its apparent speed.

By this technique, indirect as it was, astronomers were able to plumb the neighborhood of the sun. Stars were found to be more and more sparsely scattered with increasing distance; the sun was suspended at the center of a flattened cloud. This model became known as the "Kapteyn universe," named for the Dutch astronomer who performed and inspired the most important of these studies in the early part of the twentieth century. About the central cloud, a separate concentric ring of stars was thought to form the Milky Way, hanging in the background of the sky. (See plate, page 196.) This model seemed to be supported by John Herschel's discovery of a nebula with a central cloud surrounded by a ring.

The discovery of nebulae had continued after John Herschel's death, and by 1900 a puzzling fact had been added: nebulae are unevenly distributed over the sky. The large, shimmering nebulae containing stars were found almost exclusively near the band of the Milky Way, while the small spindly or round nebulae without stars unmistakably avoided the Milky Way. Some astronomers construed this to mean that the small nebulae were within our own system; others came to the opposite conclusion and suggested that there were clouds of dark matter obscuring the smaller nebulae.

During the nineteenth century, three models of the Milky Way

Galactic Plane

Our galaxy at the close of the nineteenth century. According to views then current, the galaxy consisted of an outer ring of stars and a central cloud. The sun, indicated by the cross, was placed slightly above the plane of the galaxy to account for the slight asymmetry of the Milky Way. (*Reproduced from A. S. Eddington:* Stellar Movements, *London, 1912*)

competed for favor: (1) a disk resembling the original galaxy of Kant and Herschel; (2) an unlimited slab of stars, such as suggested by William Herschel in some of his later writing; (3) a central cloud surrounded by a distant ring of stars. To the question of the *form* of the galaxy was added the question of its *uniqueness:* Is our galaxy the only object of its kind within the visible universe? Astronomers knew they could not logically separate these two questions, because the recognition of the galaxy as a unique object presupposed the failure to find anything like it in the sky; unless they could discern *its* form, astronomers could not compare it with other nebulae. Many of those who favored the ring model also felt that the galaxy was unique, and in this respect they had returned to the tradition of the Ancients. They again placed man at the center of the universe, but now they could adduce substantial evidence: they had examined the sky and found nothing else like the Milky Way within their reach.

But the true identity of the Milky Way was beyond the grasp of research at the time because there were still very few reliable measures of distance for even the nearest celestial objects, let alone the spiral nebulae and the star clouds of the Milky Way.

The times were sensitive to philosophical impressions, and man's view of his place in the eyes of God still influenced his astronomical view—at least among those who wrote popular books on astronomy. Charles Darwin, whose book *On the Origin of Species* had been published in 1858, expressed the view that man was an element in a complex organic universe, and opened the way to a scientific discussion of the possibility of life on other planets. But even Darwin referred to *the* origin and thus conveyed the strong implication of man's uniqueness. The question remained whether our planet was uniquely suited to intelligent life or whether other planets were inhabited also.

This was essentially an astronomical problem—at least it could be made astronomical if it were reduced to the simple question: How likely is it that there should be another planet somewhere in the universe at a comfortable distance from a star? If the question is put in these terms, the computation hinges on having a theory for the formation of planets, and the answer will depend on whether one accepts a theory which makes planets extraordinarily rare—such as the theory of Georges Louis de Buffon that the solar system emerged from the chaos following a near-collision of the sun with another star—or supposes that planets are a rather natural byproduct of the formation of stars. This latter view had been espoused by Laplace, and as the years rolled on, more and more astronomical evidence seemed to support its general, if not its specific, features.

But some biologists raised another question: Given a comfortable planet, what is the chance that life will spring up and evolve to the stature of man? (There is another question: Given man, how long will he permit himself to survive?) None of these questions has been answered satisfactorily.

One of the first, and still one of the most comprehensive, attempts to discuss man's claim to uniqueness was published in 1904 by Alfred Russel Wallace, who had, simultaneously with Charles Darwin, proposed natural selection as the guiding force of organic evolution. Wallace was a religious man and he was well read in astronomy. These, and the fact that he had the objectivity of a bystander to the astronomical debates, give his book *Man's Place in the Universe* an unusual interest today. He divided the argument into two parts: astronomical and biological. He conceived the Milky Way to be unique in the observable universe, although he admitted the possibility of unseen systems beyond our vision. The island universes of Kant were replaced with a single enormous continent, and man sat at the center.

He viewed the Milky Way as a dynamical system undergoing an evolution of its own; he supposed, as had Herschel, that stars were being formed within nebulae. The small nebulae without stars he thought to be young condensations of dust; with the passage of time these clouds would also contract into stars. The new stars would shine upon the remaining gas and dust and produce such objects as the Great Nebula in Orion. In this manner the absence of small nebulae near the band of the Milky Way was to be explained; those nebulae had already undergone star-birth and had vanished.

According to this conception, spiral nebulae were swirling clouds of dust and glowing gas in which future stars would be born. All nebulae were thus encompassed within a single evolutionary scheme— and all ended in the production of stars. The isolation of our central star cloud from the ring of stars forming the Milky Way was explained as a result of the cloud's collapse when it condensed into stars.

Wallace's book is a fine example of man's ability to rationalize a broad set of astronomical facts into a single, coherent scheme. He concluded that there could be a sizeable number of planets on which life *might* spring up; but he saw so many terribly unlikely steps required for the birth of life that he could not believe it had actually arisen elsewhere. Thus, Wallace concluded that man was unique in the universe; he could not see how the odds against life could be overcome more than once.

The use of probabilities was a standard argument; it was repeated by Shapley later, to the opposite conclusion—but it is a false argument.

At present, the question of life in other solar systems is essentially unanswerable because we do not know how to frame the probabilities. The calculation requires estimating whether, and how soon, a given host of atoms will combine to form the appropriate complex of molecules. Often, an arbitrary and very small figure has been chosen for the likelihood of this step in the evolution of life. But such molecules have recently been discovered in interstellar space, and the calculation must now be revised.

To argue against the probability of an inscrutable process has always been a sure path to error. Whatever may be today's view of the likelihood of life elsewhere in the universe, we can be sure that further research will not make it seem less likely. Those of us who are already convinced of the existence of life elsewhere need not fear being dissuaded; those who are uncertain may look forward to certainty.

Many aspects of Wallace's conception of the Milky Way and its development are retained today. New stars are still thought to be formed in thick clouds of dust and gas, and most of these are found near the plane of the Milky Way. The collapse of a galaxy into a system of stars from an original amorphous cloud—an idea held by the Ancients—is still a popular conception. Wallace's view that the galaxy is a dynamical system undergoing continual change receives new confirmation each year.

The anthropocentric view of the universe, as expressed by Wallace and others, was persuasive in its sweep and perhaps comforting in its implications; evidently not all men abhorred the idea that we occupy the center of the universe. But Wallace was avowedly theistic and this set him apart from many, if not all, astronomers of 1900. Most astronomers by that time either avoided sounding theistic or managed to divorce their theology from their astronomy. One man, for example, came out quite flatly against anthropocentrism. Arthur S. Eddington, who later became the greatest of the English mathematical astronomers, wrote in 1914: "By a natural reaction from the geocentric views of the Middle Ages, we are averse to placing the earth at the hub of the stellar universe, even though that distinction is shared by thousands of other bodies." The implication from this statement is that Eddington, and those who felt as he did, were relieved when the center of the Milky Way was proven to lie at a considerable distance from the sun and when the spiral nebulae were proven to be islands like our own.

The view that the Milky Way was one among a multitude of spiral nebulae had been put forth in the professional literature of astronomy by Stephen Alexander, Professor of Mathematics and Astronomy at the College of New Jersey, later Princeton University. In 1852 he suggested that the irregular form of the Milky Way might be attributed to spiral arms reaching out from our central location and seen in projection against the sky. He suggested that we are located within a central cluster containing the brighter stars of the sky, and from the fact that the brighter stars are slightly concentrated toward the plane of the Milky Way, he inferred the cluster to be spheroidal.

Alexander's discussion was rather abstract and vague; it did not attract much attention. Its central idea may have been on the minds of many astronomers at that time, but there simply was nothing that could be done to prove or disprove it; the data were lacking.

Even at the end of the nineteenth century, the most explicit discussions of the problem were couched in plausibilities because the evidence

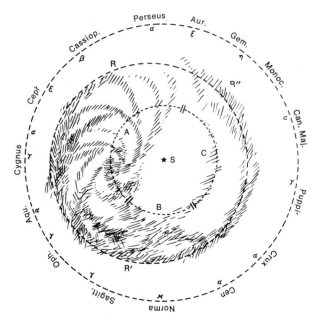

Our galaxy as a spiral nebula. This was the first published diagram depicting our galaxy as a spiral nebula. When C. Easton published it in 1900, he said that it was to be considered schematic, and further research has proven him right, although the general appearance of our galaxy is quite well represented. The sun is at the point marked "S," and the center of the spiral pattern was placed in the direction of the constellation Cygnus. This diagram is among the first to show the sun in an off-center location, and it was generally considered as an interesting, but unconvincing, speculation. (*From O. Struve and V. Zebergs:* Astronomy of the Twentieth Century, *New York, 1962—Harvard College Observatory*)

was circumstantial and uncompelling. The first diagram depicting the Milky Way as a spiral was published in 1900, by C. Easton, an astronomer working in Holland, but he added a disclaimer to the effect that the diagram was purely suggestive and should not be taken literally. (See plate, above.) Easton attacked the idea that the Milky Way was a ring of stars lying in the background; he said, "In reality [the ring model is] incompatible with the present state of our knowledge of the galactic phenomenon," and he went on to say that there was little reason to hope that the problem of the structure of the visible universe would be solved in the near future.

The incompatibility was of a rather general sort, and therefore not easily dismissed. He argued that, if one is to speak of a ring, the ring must be fairly uniform in its width and its density. Small-scale fluctuations were demanded by the irregularity of the Milky Way, and if it could be proved that the hypothetical ring were twisted and distorted, then the notion of a ring would lose much of its meaning.

Easton focused on two features of the visible Milky Way: its dramatic variation of brightness around the sky and the fact that it is widely split in several regions. He noted that it is much brighter in the constellation Aquila than in Monoceros, and that its brightness "diminishes pretty gradually from Cygnus to Casseopeia" in the northern sky, and the same sort of gradation occurs in the southern sky. The brightest portion of the Milky Way lies in Sagittarius, which is in the southern sky on summer nights.

Easton pointed out that if the putative ring were uniform and if the earth were located eccentrically, as it would have to be to account for the gradations of the Milky Way, we should see a broader, looser band of stars where we are closer to it. Each surface element of the band would have the same brightness.

But these features of a ring did not agree with the actual appearance of the Milky Way, and Easton favored a spiral pattern, whose various convolutions would produce regions of greater brightness. He placed the center of the system among the star clouds of Cygnus, and he supposed that the sun is about one-third of the way from the center to the edge.

In 1912 Easton published a photographic map of the Milky Way with the comment

> This picture, as we have it now before us, speaks for itself. It leaves next to nothing of the traditional simplicity of the galactic zone, as it is still described in textbooks: "a broad and ample road," vague and rather uniform, split in two over half its circumference. On the contrary, what strikes us most in our photographic chart is the sharp definition of many features, especially near the axis of the Milky Way, and the truly perplexing intricacy of structure, presented by the greater part of the zone.

Easton attributed the enhanced brightness of the Milky Way in the region of Sagittarius to an arm of the spiral pattern lying along the line of sight in that direction.

Some astronomers had stated that if the Milky Way were proven to be a spiral nebula, then *all* the spirals seen in the sky must be

Milky Ways, but Easton disagreed. He said, "I wish only to remark that nobody will deny the existence of a whirlpool because he sees a number of small eddies in the convolutions of the great one. I think we may safely assume that the great majority of the small spiral nebulae, if not all, form part of our galactic system."

So it seemed possible to encompass most of the nebulae into a single enormous system—our own galaxy.

Eddington summarized the dilemma in 1914 as follows:

In the days before the spectroscope had enabled us to discriminate between different kinds of nebulae, when all classes were looked upon as unresolved star-clusters, the opinion was widely held that these nebulae were "island universes," separated from our own stellar system by a vast empty space. It is now known that the irregular gaseous nebulae, such as that of Orion, are intimately related with the stars, and belong to our own system; but the hypothesis has recently been revived so far as regards the spiral nebulae. Although the same term "nebula" is used to denote the three classes—irregular, planetary, and spiral—we must not be misled into supposing that there is any close relation between these objects. All the evidence points to a wide distinction between them. . . .

It must be admitted that direct evidence is entirely lacking as to whether these bodies are within or without the stellar system. Their distribution, so different from that of all other objects, may be considered to show that they have no unity with the rest; but there are other bodies . . . which remain indifferent to galactic influence.

Eddington went on to note two alternative explanations for the peculiar and unmistakable manner in which the spirals avoided the plane of the Milky Way. "Indeed, the mere fact that spiral nebulae shun the galaxy may indicate that they are influenced by it. The alternative view is that, lying altogether outside our system, those that happen to be in low galactic latitudes are blotted out by great tracts of absorbing matter similar to those which form the dark spaces of the Milky Way." He remarked that it seemed more fruitful to assume the spirals to be external stellar systems rather than nebulae within the Milky Way.

But there were no means to measure the distances of the star clouds, the clusters, or the nebulae. Triangulation fell far short, and without distances from which to determine the absolute dimensions of the nebulae nothing further could be said.

18 SHAPLEY
AND THE STAR CLUSTERS

The key that would ultimately unlock the enigma of the nebulae was discovered by Miss Henrietta Leavitt at the Harvard College Observatory in 1908. Leavitt was a research assistant at the observatory and had focused her attention on the discovery of variable stars, the determination of their periods of variation, and the limits of their brightness. One group was of particular interest to her: variable stars in the Small Magellanic Cloud—a swarm of stars, clusters and nebulosities in the Southern Hemisphere. Harvard had installed a telescope in Peru, so Leavitt had an abundant and unique set of photographs to study.

The light of these variables showed a quick rise to maximum brightness, a slow descent, and a prolonged minimum—in intervals of one day to a month or more, depending on the star. In 1908, Leavitt drew attention to the fact that the stars of the longest period were also those of the greatest brightness, and four years later she found a detailed relationship which was undeniable: Brightness increased with period. In reporting this result, Edward C. Pickering noted that it raised a number of interesting questions "with regard to distribution, relation to star clusters and nebulae, differences in the forms of the light curves, and the extreme range of periods."

Variable stars were also known to be scattered among the star clouds of the Milky Way. These had become known as the "Cepheid variables," because the first of them had been the star *delta* in the constellation Cepheus. But these stars lay at various distances and their true brightnesses had not yet been measured.

Leavitt continued her search for variables in the Magellanic Clouds and she discovered 2400—more than the total known when she commenced working. The relationship between period and brightness among stars of the Clouds was sharpened, and it became known as the "period-luminosity relation."

In Potsdam, a young astronomer named Ejnar Hertzsprung read Pickering's announcement of the relationship found by Leavitt and he realized that he possessed enough data to determine an average of the true brightness of Cepheid variables within the Milky Way, even though individual brightnesses were still beyond grasp.

The procedure he proposed for the Cepheids was not a precise one—it required making an assumption which was known to be somewhat erroneous. But the alternative would have been to wait a decade for more precise data. Hertzsprung assumed that all of the Cepheids in the neighborhood of the sun could be treated as though they were of the same true brightness. Those that appeared brighter were assumed to be closer, so their apparent motions across the sky were enhanced by their proximity to the earth. To compensate for this purely geometrical effect Hertzsprung reduced the apparent motions of the brighter stars and increased those of the fainter stars by ratios corresponding to their presumed distances from the earth.

He then assumed that the variable stars were stationary with respect to the average of the nearby stars and that their apparent motion was just a reflection of the sun's motion among the stars. This motion had previously been determined: twelve miles per second in the direction of the constellation Lyra. Then, in much the way that a passenger in a moving auto may estimate the distance of an object in the landscape by noting how rapidly it moves past his window, the astronomer computed the distance of the "average" variable star.

He then compared the true brightness of his variables with the *apparent* brightness of Leavitt's variables of the same period in the Small Magellanic Cloud. The distance he thus determined for the Cloud was 30,000 light years. It lay just at the edge of the Milky Way.

Underlying Hertzsprung's computation was the assumption of uniformity—the most important idea in astronomy: The properties of stars near the sun are not different from the properties of the distant stars. This was the basis for his assuming that the variables near the sun could be compared *in detail* with the variable stars of the Small Magellanic Cloud. His contention was supported by the similarities between the curves of light variation displayed by the two groups of stars and the similarity in the colors of the two groups. But these similarities were far from proof.

When Hertzsprung published his distance for the Small Magellanic Cloud he caused very little stir among astronomers, because his result seemed to suggest that the Cloud was merely a fragment of the Milky Way. John Herschel had examined both the Large and the Small Magellanic Clouds when he was in South Africa, and he found them to contain such a variety of objects that he could not resist concluding that the Clouds were full-blown galaxies, separated from our own. But Herschel had no way to prove this, so the Clouds remained mere curiosities of the southern sky.

Hertzsprung's result did not change our picture, because he did not claim to have set the clouds far out in space. In fact, his publication contained a typographical error and an unfortunate guess. Both of these slips tended to make the Small Magellanic Cloud seem closer to the sun and therefore smaller and less exciting than it later proved to be. The distance he actually derived was 30,000 light years, but this result was printed as 3000 light years, making the Cloud a fairly close neighbor of the sun. Also, Hertzsprung guessed that the Cepheid variables were much redder than they later proved to be; this error had the effect of making the stars of the Cloud appear to be brighter than they were, implying that they were closer. The difficulty originated from Hertzsprung's need to compensate for the fact that the measurements of

Leavitt were made in blue light, while those of the Cepheids around the sun had been made in red light and there were no measurements with which to make an exact correction for the difference. Hertzsprung's error made him underestimate the distance by almost a factor two, so he should have published 50,000 light years as the distance. If he had, the Small Magellanic Cloud would have won the claim as the most distant object in the sky—temporarily.

He assumed that all of the Cepheid variables obeyed Leavitt's relationship between period and luminosity, so he was able to determine their individual distances. They lay in all directions about the sun; we were nearly, but not precisely, in the center of the system of variable stars. For the first time, astronomers had determined individual distances to virtually all members of a particular group of stars, and the result confirmed the earlier conclusion that the sun was located at the center of the stellar universe.

A year later, Henry Norris Russell at Princeton, working independently of Hertzsprung but starting from a similar set of premises and using similar data, came to the same conclusion. In the midst of struggling with immense tables of data, he was approached by a promising young student: Harlow Shapley.

In his reminiscences, published in 1969, Shapley recounted how he came to Princeton. Graduating from high school, he had worked as a newspaper reporter and then decided to enter the school of journalism at Missouri University in Columbia, Missouri. Having arrived with about two hundred dollars, he found that the opening of the new school had been delayed another year. He says:

> So there I was, all dressed up for a university education and nowhere to go. "I'll show them" must have been my feeling. I opened the catalogue of courses and got a further humiliation. The very first course offered was a-r-c-h-e-o-l-o-g-y, and I couldn't pronounce it! (Though I did know roughly what it was about.)
>
> I turned over a page and saw a-s-t-r-o-n-o-m-y; I could pronounce that—and here I am!
>
> From then on things went swimmingly—I had found my field.

Shapley received his bachelor's degree from Missouri University in 1910 and went to Princeton, where he concentrated on eclipsing bi-

Schematic diagram of a pulsating star. Variations of size and temperature cause the apparent brightness of this star to vary periodically.

naries and developed some new ideas "on getting the distances of eclipsing binaries from studies of their colors and spectra—a forerunner of the theories on Cepheid variable stars."

This remark, also from his reminiscences of 1969, suggests a close relationship between the roles of eclipsing variables and Cepheid variables in Shapley's work—a relationship which was due in part to the remaining uncertainties concerning the nature of these stars and the difficulty of distinguishing between them.

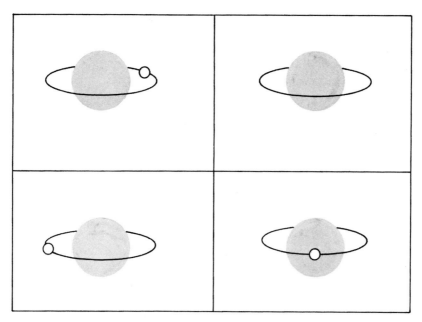

Schematic diagram of an eclipsing variable. If two stars are in orbits about each other, and if they are viewed from the right direction, they will alternately block each other's light. In this diagram, the small star is assumed to be hotter than the large star. When the hotter star is eclipsed, the system will become substantially darker; when the cool star is eclipsed, the system may only become slightly darker.

Through his work on eclipsing variables Shapley came to realize the possibility of determining the distances of individual stars without relying on triangulation, a technique which was confined to the immediate neighborhood of the sun. Even a very distant eclipsing star can be made to divulge the dimensions and true brightnesses of its members; the distance may then be computed from the apparent brightness. The technique is the following.

Imagine two stars in orbits about a common center; both orbits will lie in a single plane, and if the earth also lies in this plane each star will periodically block the light of the other. If the two stars are very close together—for example, separated by a distance equal to their diameter—the eclipses will be visible from a fairly wide band of directions near the plane of the orbit. Such "eclipsing variables" would look exactly like a single star in a telescope, and their light would remain nearly constant except during the interval of the eclipse, when it would fall abruptly.

Suppose one star is very much smaller than the other, but that it is hotter and its total brightness is the same. When the small star vanishes behind the large star, the total light will appear to be cut in half. This will occur on each orbit, and the other eclipse—when the small star slides across the face of the large star—will be very shallow, perhaps invisible except with the most careful measurements. Suppose, further, that the interval between the deep eclipses is ten days while the eclipse itself lasts one day. From this we would know that the large star had a diameter which is just one-tenth of the circumference of its orbit about the small star—assuming that the orbit is a circle.

The circumference of the orbit may be determined by measuring the velocity of the stars about each other with the spectroscope, and multiplying by the orbital period. From this, we determine the true diameter of the large star, and its total surface area.

Next we need to derive the brightness of each element of the star's surface, say each square meter. This can be done by examining the color of the star and employing a mathematical relationship known to describe the brightness fairly well in terms of color. Multiplying the brightness of each square meter by the total area of the star, we find the total brightness of the star. Comparison with the apparent brightness gives the distance.

This is a simplification of the work Shapley did for his thesis; the actual work was complicated by such factors as the ellipticity of the orbits, the uneven distribution of light on the face of the stars, the fact that the stars did not pass directly in front of one another, and so on.

Shapley says those days at Princeton were happy days because it was obvious that they had found a gold mine in the eclipsing variables, and "the field had not been worked very much." The publication of Shapley's thesis in 1913 brought him to the attention of astronomers all over the world.

Eight miles by pack trail from Pasadena, California, stands Mount Wilson—a mile above the flat valley of Los Angeles. On that peak, George Ellery Hale constructed a solar observatory in 1904; in 1907, a team of mules dragged a glass disk sixty inches in diameter up the trail to the summit. This was to be the light-collector for the largest telescope in the world, and was to be used for the photographic study of starlight.

By the close of the nineteenth century, photography had revolutionized astronomical measurement, giving precision, permanency, and plenitude almost beyond the imagination of those who had worked with their eyes peering through brass telescopes at dancing images. With the sixty-inch telescope, the Mount Wilson Observatory soon gave promise of becoming the astronomers' Mecca. The scientific staff grew rapidly and shared its collection of photographs with visitors from all over the world.

Hale heard of Shapley and arranged an interview; they met in a New York hotel and talked of the opera and the theater, much to Shapley's dismay. Within a short time, however, Shapley had been offered a job at Mount Wilson.

His first project was to take a close look at Cepheid variation and attempt to determine its cause. In his mind was a model of a pulsating star: a single star whose surface surges outward and inward like the surf, carrying waves of heat and light to outer space. Without a precise theory for this pulsation, he could not predict the light variations of such a star, but even the most elementary theories permitted one important statement: The period of pulsation would depend only on the density of the star; in fact, the product of the star's pulsation period and the square root of its density would be the same for all stars. In other words, a star less dense than our sun would pulsate with a longer period—a decrease by four in the density would imply an increase by a factor two in the period.

(If the sun were a pulsating star its period would be about one hour; it does not pulsate with this period, because its predilection to pulsation is overcome by internal resistance.)

Viewed close at hand the eclipsing star looks totally different from the pulsating star, but from a distance the distinction is not so clear, particularly if the members of an eclipsing pair are very close together and the light is changing nearly all the time. Shapley showed that most of the Cepheid variables must be pulsating rather than eclipsing stars because the assumption that they were eclipsing led to the absurd result that each star was inside the other. They would have to be closer together than the total of their radii. (It seemed absurd at the time, but the idea was again advocated forty years later—and then again discarded.) The conclusion that Cepheid variables were pulsating gave direction to theoretical studies, and it eliminated fruitless possibilities. Shapley indicates the following of its effect on his own work:

> After I had published the paper I worried whether I might have taken the theory from Russell. He and I had discussed it. I claimed that I had taken it from him, and I was going to write an apology. It seemed preposterous that a professor would have a student who would snitch his good ideas. But Russell said he had never heard of the theory, and anyway he was doubtful that it was correct. But it was; it stands up—it is, in fact, basic to astrophysics.
>
> All this made me confident that I could do something significant at Mount Wilson if the people there gave me a chance. And they did give me a chance. Working on eclipsing stars for two or three years made me grow in maturity rather rapidly. I realized that I could do things other people could not or would not do, and therefore I was useful.

On his way to Mount Wilson, Shapley had stopped at the Harvard Observatory, where Solon Bailey said to him, "When you get there, why don't you use the big telescope to make measures of stars in globular clusters?" His time was not entirely his own, however, as he had been assigned to help Frederick Seares with observations of the colors and magnitudes of stars in the Milky Way. The work was tedious, but it familiarized Shapley with the sixty-inch telescope and with observational techniques that later proved invaluable in his studies of globular clusters. He tells: "My own research at Mount Wilson was concerned almost from the first with the distances of Cepheid variables. Some of the Cepheid variables are in globular clusters, and that also interested me because the distances [of the clusters] we could get with eclipsing binaries seemed to tie up with the distances from these Cepheid stars."

Shapley determined the distances to the globular cluster variables in the manner of Hertzsprung's earlier work, but he pushed the analysis one stage further and attempted to prove that the Cepheids in the neighborhood of the sun—the stars studied by Hertzsprung and Russell—obeyed the period-luminosity relationship of Leavitt. If he could prove this, it would be support for his contention that the globular cluster members obeyed the same relationship. His graphs indeed showed that the bright Cepheids were those of longer period. All looked well for his argument, so he assumed that the three types of variable stars (in the Small Magellanic Cloud, in globular clusters, in the neighborhood of the sun) were identical.

This assumption permitted him to compute the true brightnesses of the cluster variables, and from the apparent brightness and the assumption that there was no obscuration in space he found the distances of the clusters. There were a dozen clusters with variables, and Shapley looked for ways of measuring the clusters that contained no variables. He found that the brightest stars in a cluster were about three times as bright as the variable stars, so even if variables had not been discovered in a particular cluster Shapley could compute how bright they would have been, and from this could estimate the distance. Similarly, the diameters of clusters appeared to be fairly uniform, so he could estimate the distance from the apparent size. In this way he compiled the distances of sixty-nine clusters, and when he plotted their positions he found that they formed a slightly flattened system in which the sun was decidedly peripheral.

The average globular cluster was much farther from the sun than the Cepheids measured by Hertzsprung, and their collective center was well beyond the limits of the local star cloud. Yet they showed a tendency, as did the spiral nebulae, to avoid the plane of the Milky Way.

Shapley then assumed that the system of globular clusters defined the center of our galaxy; this put the sun well away from the center,

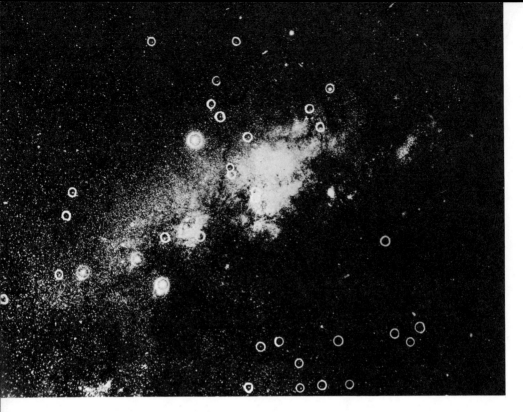

Globular star clusters in Sagittarius. Shapley suggested that faint clusters, indicated by the circles, comprise a spherical swarm about the center of our galaxy.

and the diameter of the galaxy was expanded by a factor ten over the estimates then current. Today the most intriguing question is: "What led Shapley to assume that *this* group of objects should be considered to define the center of the galaxy?"

His own writing implies that he had no alternative, but this seems to oversimplify the matter. A half-dozen lines of study had consistently indicated that the sun was near the center of the galaxy and that the entire galaxy was hardly more than 30,000 light years in diameter. The globular clusters were the *only* objects which contradicted the picture, and they did it flagrantly, suggesting a diameter of 300,000 light years. Shapley chose to center his galaxy on the exceptions; controversy followed.

And yet, despite the exceptional nature of the globular clusters, astronomers who disagreed with Shapley did not deny that the clusters were galactically distributed nor did they deny that the *relative* dis-

The 60-inch reflector of the Mount Wilson Observatory. This is the instrument with which Shapley carried out his search for the center of the Milky Way. (*Hale Observatories*)

tances assigned to individual clusters were correct; they merely refused to accept the absolute distances. As Shapley put it, if the distance to *one* cluster could be proven correct, then the other distances would follow and the center of the galaxy would be located.

Astronomers were quite ready to accept the peripheral location of the sun; what they could not accept was the enormous scale of the galaxy. One reason for this reluctance was that the size derived by Shapley spoke strongly against the island-universe concept of the spiral nebulae. Shapley himself said that the spirals were not true galaxies, and at various times he suggested that they were members of our own system, or that they were spiraling gas clouds lying in the neighborhood of our galaxy.

He advanced two arguments for believing the spirals to be smaller than our galaxy. A colleague at Mount Wilson, Adrian van Maanen, had detected the rotation of several spirals, and if the spirals were external galaxies, the apparent motions would require rotational velocities far in excess of the speed of light. If these observations were valid, the spirals had to be within the Milky Way.

Shapley's other argument was based on the appearance of a nova in the Andromeda Nebula, M31. In 1885, a star had appeared near the center of the spiral, becoming as bright as the entire nebula. It had been pointed out that if this nova was assumed to be as bright as novae within the Milky Way—several of whose distances were known—this would put the spirals well within our own system. Also the conception that a single star could become as bright as an entire galaxy seemed outlandish; this also argued that the spirals were much smaller than our galaxy.

Consistent objections to Shapley's work came from Heber D. Curtis at the Lick Observatory. He objected to Shapley's use of the Cepheids and insisted that the size of the galaxy had been grossly overestimated.

In 1920, the two men were brought together for a public discussion by the National Academy of Sciences. Historians have since labeled this "The Great Debate," and it was nearly unique in American science, but Shapley's 1969 reminiscences contain remarks which suggest that he held a rather different attitude at the time:

As for the actual "debate," I must point out that I had forgotten about the whole thing long ago . . . Then, beginning about eight or ten years ago, it was talked about again. To have it come up suddenly as an issue, and as something historic, was a surprise, for at the time I had just taken it for granted. . . . I don't think the word "debate" was used at the time. Actually it was a sort of symposium, a paper by Curtis and a paper by me, and a rebuttal apiece.

On the platform, Shapley summarized his earlier arguments for the enormous distances of the globular clusters and concluded with a comment that the island-universe theory of spirals "probably stands or falls with the hypothesis of a small galactic system." If the galaxy should prove to be as large as Shapley claimed, the spirals were likely to be engulfed by it.

Curtis insisted that the spirals were comparable to our own galaxy. He claimed that the determination of true brightness for the Cepheids was far less certain than Shapley's discussion had suggested. New data, and other data that Shapley had not used, were extremely discordant, so he felt that the "available observational data lend little support to the fact of a period-luminosity relation among galactic Cepheids . . . it would seem wiser to wait."

Concerning the novae, Curtis pointed out that "a division into two . . . classes is not impossible"; in other words, the novae observed in spirals might be brighter than ordinary novae observed within the Milky Way. If so, the spirals might be external galaxies.

The symposium settled nothing. Only new data could do that. In his reminiscences, Shapley recalls,

Anyway, it was a pleasant meeting, and our subject matter was the scale of the universe. That was what I was prepared to talk about and did talk about, and I think I won the "debate" from the standpoint of the assigned subject matter. I was right and Curtis was wrong on the main point—the scale, the size. It is a big universe, and he viewed it as a small one. From the beginning Curtis picked on another matter: Are the spiral galaxies inside our system or outside? He said that they are outside our system. I said, "I don't know what they are, but according to certain evidences they are not outside."

But that was not the assigned subject. Curtis, having set up this straw man, won on that. I was wrong because I was banking on van Maanen's measures of motions in spirals. If you have large [apparent] motions you are dealing with things near at hand. I consider this as a blunder of mine because I faithfully went along with my friend van Maanen and *he* was wrong on the [apparent] motions of galaxies—

that is, their cross motions. Although Curtis and Hubble and some others discredited van Maanen's measures and questioned his conclusions, I stood by van Maanen.

It soon became evident that both men had won and lost the debate. Knut Lundmark, a Swedish astronomer, showed that there *were* two kinds of nova: one is the common kind which flares to a brightness of 10,000 times the sun; and the other—the supernova—which can become 10,000 times brighter even than *that*. Supernovae do in fact become as bright as the entire Milky Way, and it is supposed that Tycho's nova of 1572 was a supernova. When this distinction had been clarified, the main argument against the island-universe concept vanished. Later, van Maanen's measures were proven incorrect and another of Shapley's arguments for the small size of spiral nebulae vanished.

On the other hand, Shapley's use of the Cepheids to measure the distances to the clusters was verified, and became the most important astronomical yardstick.

In 1921, Shapley left Mount Wilson to take on the role of Director of the Harvard College Observatory, one of the most prestigious astronomical posts in the country. His acceptance of the post came as a surprise to his mentor Russell, who Shapley quotes as saying, "Oh, no, no, you wouldn't want to do that." Another astronomer wrote to a friend that he didn't think "Shapley would be so foolish as to give up his astronomical career just to be director of that observatory."

Shapley's departure from the West Coast took him away from the great telescopes, and it is difficult to avoid thinking that his decision must have been partly a reaction to his disappointment at events that had brought him so close to recognizing the spirals as external galaxies, and which had then withheld the discovery from him. In support of this view are Shapley's protesting that Curtis had set up a "straw man" during the debate and his later interpretation of his own work, in which he gave the impression of having brought man through a revolution by taking him away from the center of the universe. Historically, this is inaccurate: by 1920, man had already unseated himself—although it was left to Shapley to locate the chair.

But I think another interpretation of Shapley's move and his subsequent life is more satisfactory. He had been interviewed for the Harvard job *before* the spiral nebulae had been shown to be external galaxies, and Shapley was an ambitious young man. When he measured the distance to the center of the galaxy, he must have wondered what he could possibly do next to match this achievement. The question would have been particularly difficult for him to answer because he thought that the spirals were secondary creatures, and to measure them would be mere dénouement.

So he became a public man: he administered his observatory; he worked to bring foreign scientists to America; he helped organize UNESCO. In short, he did what John Herschel, Isaac Newton, and many others had done—he reached beyond science.

The current view of our galaxy. In this schematic side view, the sun lies to the left and the globular star clusters are indicated by the large dots. The dark rift along the principal plane represents the thin layer of dust which obscures starlight and the light from external galaxies. The diameter of this system is now estimated at 75,000 light years. This conception is influenced by our view of external galaxies; for example, see the plate on page 236. (*Yerkes Observatory, University of Chicago*)

As a graduate student at Harvard, I came to know Shapley. Two dozen students met in his home every full moon and congregated in a large room where we found puzzles, newspaper clippings, new books and magazines spread out on a long table. For half an hour, we amused ourselves while he walked about, watching, injecting a joke, or showing a stymied student how to complete a puzzle. Then we sat about the table and for an hour or two danced to his tunes as he led us through a series of intellectual minuets.

I have never seen a quicker mind, a more agile sense of humor, or a more complete absence of what usually passes for humility. He needled and stimulated us; he told us things we should have known or that we knew we would never need to know; he told us about his discoveries of ants in the Kremlin, where he found a pair on the table-cloth during a state dinner and preserved them in Vodka, and in India, where he told Mr. Nehru that he would find a certain species of ant under a particular rock—and did. Shapley's irreverence often had an unexpected effect. He spoke of Albert Einstein as "Saint Albert" during an after-dinner speech that I attended, and he made me wonder if perhaps I hadn't revered Einstein a bit too much. And he claimed to have given Robert Frost the idea for his cosmological poem "Fire and Ice." I am willing to believe that he did, because I like to imagine those two men in a room together.

19 INTO THE REALM
OF THE NEBULAE

✲

On August 28, 1907, as the mirror of the sixty-inch telescope was being trundled up the trail to Mount Wilson, Hale was notified that the disk of glass for a one-hundred-inch mirror had been cast in France and would be sent to California as he had requested. There was no money for the completion of the one-hundred-inch telescope, but Hale had succeeded in stimulating a wealthy friend, J. D. Hooker, to give the first $45,000 toward the mirror.

The arrival of the one-hundred-inch mirror in California brought bitter disappointment because there were layers of bubbles throughout the disk, and the glass from the separate melting pots appeared to have been incompletely fused. Frustrated, Hale ordered the disk put into dead storage and told the factory to start over. In 1910, after the construction of a new annealing oven, the second attempt was made, but on this occasion the disaster came quickly: the disk broke as it was cooling.

At about this time Hale suffered his first of several nervous breakdowns; the money from the original gift had been exhausted and the donor had died. The project to build a greater telescope seemed stillborn until the original disk was re-examined and Hale permitted the opticians to start grinding and polishing it. All went well until the final stages, when the time came to measure the curvature of the surface so that corrections could be achieved during the subsequent perfection of the polish. The great disk was set on its edge and light from a small lamp was reflected from its wet surface and returned to the edge of a razor blade, where most of it was caught. If the mirror were a perfect sphere—an intermediate goal in producing the desired parabola—all of the light would strike the razor: deviations from perfection would send rays around the edge and they could be detected with a small telescope.

The optician found that the mirror appeared to be bent across the middle, as though it were buckling under its own weight. He turned the mirror around its axis, and again it seemed to buckle. Some of the staff were perplexed because they thought the mirror much too stiff to buckle. No one was sure, but they all agreed that if the mirror were as weak as it appeared to be it would have to be junked.

Then the incredible cause of the difficulty was located. The air in the measuring room was denser near the bottom edge of the mirror than near the top edge, and although this difference had always been trivial for smaller telescopes, it was not the case for the sixty-inch mirror. The air had bent the light from the lamp, giving the appearance of a bend in the mirror.

Once this technical difficulty had been eliminated, work progressed more or less routinely, but the pressures from this and other projects again took their toll on Hale's health. He was ordered to rest, so he took a trip to Europe and Egypt, and while traveling he heard that Andrew Carnegie had given $10,000,000 to the Carnegie Foundation with the clear implication that substantial money would be funneled toward the Mount Wilson Observatory. But even this news seemed to exacerbate Hale's nervous condition—"brain congestion," in the words of one acquaintance—and, after returning to the United States and encountering a number of difficulties, including the illness of his wife and hostilities and epilepsy within his staff, he permitted himself to be admitted to a sanitarium for another rest.

He soon recovered sufficiently to resume his life of research, travel,

The 100-inch telescope of the Mount Wilson Observatory. This is the instrument with which Hubble carried out most of his research in the "realm of the nebulae." It was completed in 1917 and is still in continuous operation. (*Hale Observatories*)

administration, and organization. By 1913, the steelwork to sustain the one-hundred-inch mirror had been cast at a shipyard in Massachusetts, sent through the Panama Canal, and placed on concrete piers atop the peak. By the spring of 1917, the great mirror was nearly completed and Hale was actively recruiting staff members.

On April 6, 1917, the United States entered World War I. Shortly after, Hale received a telegram from Edwin Hubble, a young man he had invited to join his staff: "Regret cannot accept your invitation. Am off to war."

Hubble was then twenty-seven years old; he had already turned away from the chance to become a professional boxer, had spent three years in England as a Rhodes Scholar, had practiced law for a year, and then had earned a Ph.D. in astronomy at the University of Chicago.

In at least one respect Hubble's career recalls that of the elder Herschel: both men had become professionals in another field and had then shifted to professional status as astronomers.

Hubble was the fifth in a firmly disciplined family of seven children. His father worked for a Chicago insurance firm, and the children were expected to earn money for themselves. Born in 1889, young Edwin seems to have developed an early interest in astronomy, because at the age of twelve he replied to an astronomical question from his grandfather so cleverly that the letter was published in a newspaper. The boy must have been pleased.

In high school Edwin was a quick youth, mentally and physically. The principal is reported to have handed him a scholarship to the University of Chicago with the remark, "I have never seen you study for ten minutes." He received college letters for boxing, track, and basketball; in the Army he drove a motorcycle. Hubble seems to have relished physical danger—or at least so we infer from his admission that he enjoyed army life and from stories of his escapades.

Entering college in 1906, he tutored and he worked in the laboratory of Robert Millikan, who later became one of the greatest of American physicists, and although this experience gave him a taste of laboratory research, it did not turn him to the study of physics. Instead, he accepted a Rhodes Scholarship and went to Oxford, where he read Roman and English law and learned enough to pass the bar exam upon his return to the United States in 1913.

After one year of practice, Hubble says that he decided to "chuck the law for astronomy," and that even if he was a second- or third-rate astronomer, "it was astronomy that mattered." He applied to his alma mater, the University of Chicago, and entered the graduate course in astronomy, working at the Yerkes Observatory and writing a thesis based on photographs of faint nebulae. The following excerpt from the introduction of his thesis expresses the enigma of nebulae as Hubble saw it at the time he finished graduate study and joined the Army in 1917:

Extremely little is known of the nature of nebulae, and no significant classification has yet been suggested; not even a precise definition has been formulated. The essential features are that they are situated out-side our solar system, that they present sensible surfaces, and that they should be unresolved into separate stars. . . . Some at least of the great diffuse nebulosities, connected as they are with even naked eye stars, lie within our stellar system; while others, the great spirals, with their enormous . . . velocities [toward and away from the earth] and insensible proper motions [across the sky], apparently lie outside our system. The planetaries, gaseous, but well defined, are probably within our sidereal system, but at vast distances from the earth.

In addition to these classes are the numberless small, faint nebulae, vague markings on the photographic plate, whose very forms are in-distinct. They may be planetaries or spirals, or they may belong to a different class entirely. They may even be clusters and not nebulae at all. These questions await their answers for instruments more power-ful than those we now possess.

Hubble's doctoral thesis did not solve the problem, but it seems to have strengthened his suspicion that the small nebulae were outside our own galaxy, and it certainly whetted his appetite for research with the tele-scopes of Mount Wilson.

Shortly after Hubble enlisted in the Army, the one-hundred-inch was completed; the mirror went up the mountain in July 1917. At the time, Hale was in Washington, D.C., assisting in the coordination of scientific research as the chairman of the newly formed National Re-search Council. By November, the new telescope was ready for its trial, and Hale was on the mountain with Walter Adams, of the staff, and the poet Alfred Noyes, whose brother Arthur had met Hale as a student at M.I.T. and worked with him in Washington. Hale's biographer, Helen Wright, describes the fateful night in these terms:

Hale, with Adams, climbed the long flight of narrow black iron steps to the observing platform. On the floor below, where a dim red light glowed, the night assistant pushed the control buttons. . . . The ob-serving platform rose and turned; the dome, its slit open wide to the star-filled sky, revolved in the opposite direction; the telescope itself turned until it pointed to the brilliant planet Jupiter.

As soon as the telescope was set on Jupiter, Hale crouched down to look through the eyepiece, desperately eager to know if all the years of effort had been successful. He looked and said nothing; only the expression on his face told of the horror he felt. Adams followed. His expression was a mirror of Hale's. They were appalled by what they had seen. Instead of a single image, six or seven overlapping images filled the eyepiece.

The mirror appeared to have been distorted into a half-dozen flattened areas, each producing an image of the planet. There was nothing to do but wait and see whether, in the cool evening air, the mirror would assume its proper shape. The men were hopeful because they knew that some workmen had left the dome open during the day and the sunlight might have heated the mirror.

During the evening Hale chided Noyes, saying that he should write of the war for knowledge rather than the wars of men against men. Later, Noyes rose to the challenge and wrote an epic poem, *Watchers of the Sky,* in which he referred to the first night with the one-hundred-inch telescope,

> *Up there, I knew*
> *The explorers of the sky, the pioneers*
> *Of science, now made ready to attack*
> *That darkness once again, and win new worlds . . .*
> *For more than twenty years,*
> *They had thought and planned and worked.*

But, points out Wright, Noyes "fails to record the real drama of that night in the dome. He does not describe the dark hours after the first appalling observation." After a while, they walked outside and gazed down at the lights twinkling in the valley below.

Agreeing to meet three hours later, they went to bed. Hale lay down without undressing, but he could not sleep. An hour later he got up and tried to read a detective story. But this too failed. At 2:30 A.M. he returned to the 100-inch dome. Before long Adams arrived and confessed that he too had found sleep impossible. Once again they climbed the long flight of steps to the dome floor, then the narrow flight to the observing platform.

By this time, Jupiter was out of reach in the west. They swung the great telescope over to the brilliant blue star Vega. Almost afraid to look, Hale again crouched down and looked into the eyepiece. He let out a yell. The yell told Adams all he wanted to know.

The telescope was an unqualified success. "All the agony had not been wasted."

During the summer of 1919, Hubble returned from France and went straight to the job offered by Hale. At first he concentrated on studies of nebulae within our galaxy. He suggested that the light of diffuse nebulae—the Great Nebula in Orion is the finest example—is starlight that in some nebulae has been reflected by solid particles, and in other nebulae has been absorbed and re-emitted by individual atoms. Whether a nebula is particulate or gaseous would depend, according to his suggestion, on its density and on the temperature of the embedded stars. This conjecture was later confirmed. Stars hotter than about 20,000°C vaporize the matter near them, leaving only gas; cooler stars may be surrounded by dust. Dust will partially reflect starlight without altering its spectrum.

Thus, the spectrograph can distinguish one type of nebula from the other because the light of a dusty nebula shows precisely the same features as the light of the stars shining upon it, while the light of a gaseous nebula shows distinctive features—for example, lines in the spectrum—which are unrelated to the details of the spectrum of the star and are specific to the gas in the nebula. As an indication of the process by which the light is sent out from these types of nebulae, the one type are called "reflection nebulae" and the other "emission nebulae."

The crucial point of Hubble's discovery was that the emission nebulae are invariably associated with blue, very bright stars; the nebulae could in fact be used as tracers of the blue stars. Thus, when Hubble found a stellar system with faint emission nebulae, he could immediately guess that a further search would reveal blue stars, and the apparent faintness of these blue stars would be an indication of the great distance of the system.

General interest in studies of the spiral nebulae was spurred by a chance discovery. G. W. Ritchey, the optician who had helped Hale to resurrect the imperfect disk of the one-hundred-inch mirror, found a nova on a photograph of N.G.C. 6946 in 1917, and his examination of earlier plates revealed two novae that had gone unnoticed in the Andromeda Nebula, M31. When he announced the discovery, H. D. Curtis looked through the old photographs at the Lick Observatory and discovered three more novae in two other spiral nebulae. The discovery of these novae suggested that the spirals were stellar systems and that their stellar content might be available for study.

The Swedish astronomer Lundmark, while visiting the Mount Wilson Observatory, obtained spectra of some of the bright objects in the spiral M33, and these confirmed the presumption that some of the objects were indeed stars. Then J. C. Duncan, Professor of Astronomy at Wellesley College and a regular visitor to the observatory, discovered several variable stars in M31, and it became clear that stars had been discovered in spirals.

At the Mount Wilson Observatory, a telescopic camera with a ten-inch lens surveyed the sky for objects warranting further study with the large reflectors. The anomalous character of N.G.C. 6822, an irregular nebula discovered in 1886 by the American astronomer E. E. Barnard, attracted attention. Barnard had remarked that over the course of a year the nebula seemed to have grown in size: "Probably this is a variable nebula." But these observations were made without the aid of photography, and when he discussed them in 1925 Hubble concluded that the apparent variations had been an illusion produced by the use of various powers of magnification. The nebula attracted attention for another reason: it was remarkably similar to the Small Magellanic Cloud, except that it was much smaller. In 1922 the British astronomer C. D. Perrine wrote: "This nebula . . . has no counterpart in my experience among the smaller nebulae but resembles the Magellanic Clouds." He insisted that the object was not an ordinary knot of the Milky Way, because its content was more richly varied and its structure was more complex. He suggested that it was in the same class as the Magellanic Clouds; the implication of this remark was that N.G.C. 6822 might be a galaxy detached from our own.

Duncan obtained several photographs confirming that this was a "remarkable group of stars and nebulae," and two years later Hubble began a detailed study, evidently because N.G.C. 6822 promised to combine his interest in nebulous stars with his interest in galaxies outside our own. He obtained fifty photographs with the large reflectors during the next two years; he searched them for variable stars, examined five diffuse nebulae embedded within the Cloud, and counted the numbers of stars of each brightness seen on its face.

Hubble found fifteen variables. Judging by the shapes of their light curves, eleven of them were Cepheids. He assumed that they were similar to the Cepheids studied by Shapley. If this hypothesis—he called it the assumption of "uniformity"—should lead to contradictory results then it would have to be dropped, but until that time Hubble intended to use it as the basis for measuring the distances of galaxies and star clouds containing Cepheids.

Hubble studied N.G.C. 6822 to see whether the assumption of uniformity led to self-consistent results. Several tests were possible. First, the Cepheids displayed the familiar relationship between period and apparent brightness. Using the true brightnesses determined by Shapley for Cepheids near the sun, Hubble found a distance of 700,000 light years—by far the greatest distance measured until that time. In fact, Hubble says that N.G.C. 6822 was the "first object definitely assigned to a region outside the galactic system."

Hubble then inquired what this distance would imply about other features of the cloud, particularly its relationship to the Magellanic Clouds. Shapley had looked at this question and, in the words of Hubble: "He boldly assumed an analogy with the Magellanic Clouds, and, comparing angular dimensions, size, and luminosity of the diffuse nebulae, and estimated magnitudes of the brightest stars involved, arrived at a distance 'on the order of a million light years.' This figure is comparable with that derived from the Cepheids and represents a brilliant application of the general principle of the uniformity of nature."

Hubble also found confirmation of this distance in another quarter. He counted the bright stars of N.G.C. 6822 and plotted their number against their apparent brightness, finding a curve which was similar to the previously counted population of stars in the neighborhood of the sun.

All available criteria agreed in identifying N.G.C. 6822 as "an isolated system of stars and nebulae of the same type as the Magellanic Clouds, although somewhat smaller and much more distant." With evident satisfaction, Hubble went on to say:

A consistent structure is thus reared on the foundation of the Cepheid criterion. . . . The principle of the uniformity of nature thus seems to rule undisturbed in this remote region of space. This principle is the fundamental assumption in all extrapolations beyond the limits of known and observable data, and speculations which follow its guide are legitimate until they become self-contradictory. . . . The Cepheid criterion . . . seems to offer the means of exploring extra-galactic

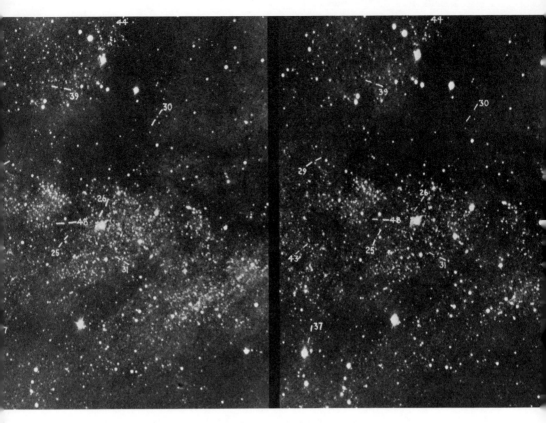

Variable stars in the Andromeda Nebula. Variable stars are indicated, and the difficulty of discovering them will be apparent to the reader. Discovery is greatly facilitated by a device known as a "blink microscope," which permits alternative viewing of two plates taken at different times. (*Reproduced from Edwin Hubble:* The Realm of the Nebulae, *New Haven, 1936*)

space; N.G.C. 6822 furnishes a critical test of its value for so ambitious an undertaking and the results are definitely in its favor.

Thus Hubble announced his intention to explore extragalactic space, using the criterion of distance which Shapley had used within the Milky Way: the brightness of Cepheid variables.

He pressed the search for Cepheid variables in M33, a spiral nebula whose brightness was second only to that of the nebula in Andromeda, M31. He found thirty-five Cepheids; they displayed a definite period-luminosity relationship, and by assuming that they were identical to the variables of the Magellanic Clouds, Hubble showed that the distance of M33 must be eight times that of the clouds. This put it 800,000 light years from the sun, far beyond the limits even of Shapley's model for the Milky Way. Examination of other criteria of distance—the brightest stars, the novae, and the small nebulosities—gave consistent results and again seemed to confirm the assumption of uniformity.

Hubble's announcement caused a sensation, and it abruptly ended the debate over the nature of the spiral nebulae. They were obviously island universes external to our Milky Way. Galaxies had been discovered.

Two years later Hubble published a detailed analysis of the Andromeda Nebula, showing that it was similar to M33. But both systems seemed to be only one-tenth the size of Shapley's model for our own galaxy.

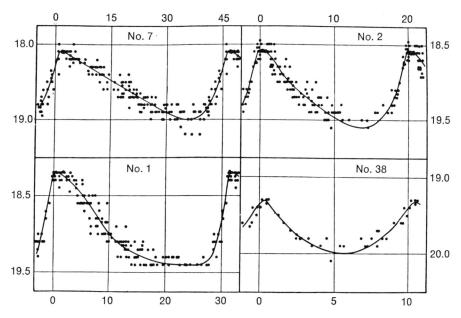

Light variations of Cepheid variables. Measurements of four variable stars in the Andromeda Nebula are plotted as a function of time measured in days. The period of variation is related to the absolute brightness of such a star. Once the relationship had been discovered for stars within our own galaxy, it was used by Hubble to determine the distances of these stars—and hence of the Andromeda Nebula. (*Reproduced from Edwin Hubble: The Realm of the Nebulae, New Haven, 1936*)

Many astronomers hesitated to identify our galaxy with such spirals. Some felt that although we might be located near the center of a typical spiral, the globular clusters and the Magellanic Clouds were too far away to be considered part of the same spiral. We seemed to be nestled among an array of galaxies: the irregular Magellanic Clouds; a local cloud which was, perhaps, a spiral; and another object whose distant center lay among the swarm of globular clusters but whose form was unknown. Shapley called this collection the "inner metagalaxy" to indicate that it was something more than a single galaxy, and he described the Milky Way as a swarm of "half-digested" star clouds.

Hubble's discovery of the island universes had not completely clarified the nature of our own system. The missing clue was the globular clusters; none had been found in either of the spirals studied by Hubble.

In the interval between the publication of his two papers announcing M33 and M31 to be external galaxies, Hubble completed a general survey of the extragalactic nebulae and succeeded in classifying them according to a fairly simple scheme. Only a few per cent of galaxies could not be classified; they were strange distortions of the normal galaxies, and Hubble paid little attention to them.

In his introduction to the classification scheme Hubble pointed to the fundamental distinction between the galactic and the extragalactic nebulae. He said: "The relationship is not generic; it is rather that of the part to the whole. Galactic nebulae are clouds of dust and gas mingled with the stars of a particular stellar system; extra-galactic nebulae, at least the most conspicuous of them, are now recognized as systems complete in themselves, and often incorporate clouds of galactic nebulosity as component parts of their organization."

Among the galactic nebulae, Hubble distinguished two types: planetaries and diffuse nebulae. The planetaries are round and well-defined, often in the form of a ring or a series of rings, with a darker central portion. In many, a faint blue star is located at the center, providing the energy with which the nebula shines. The gas of all such nebulae expands from the central star and it is escaping into space, presumably from a mild eruption or explosion on the stellar surface. The amount of matter in a planetary nebula is a minute fraction of the mass of an ordinary star.

Some of the diffuse galactic nebulae were predominantly luminous,

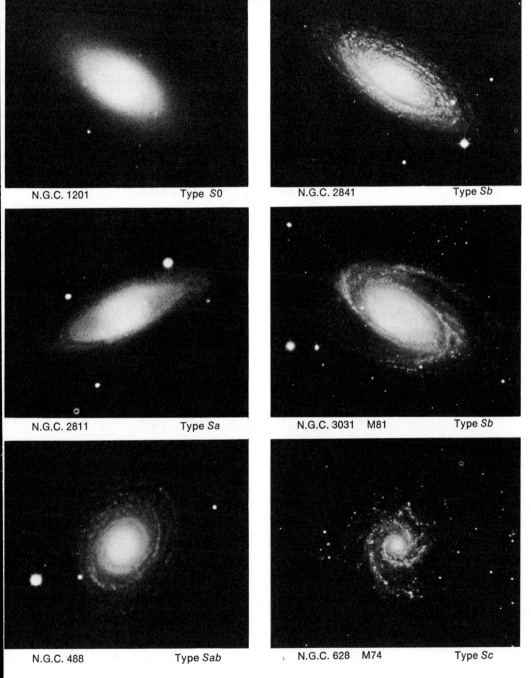

N.G.C. 1201   Type *S0*

N.G.C. 2841   Type *Sb*

N.G.C. 2811   Type *Sa*

N.G.C. 3031   M81   Type *Sb*

N.G.C. 488   Type *Sab*

N.G.C. 628   M74   Type *Sc*

Nebular types. These are external galaxies, and are classified as "normal spirals" according to the scheme developed by Hubble. (See plate, page 232.) Type *Sc* has the smallest central region and the most loosely wound and fragmented spiral pattern. Type *S0* shows no arms, but it appears to have a disk in addition to the central ellipsoid. (*Hale Observatories*)

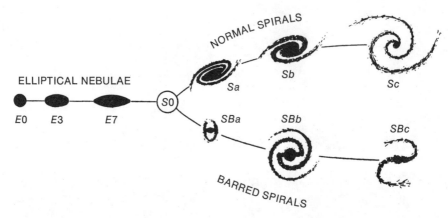

ELLIPTICAL NEBULAE

E0   E3   E7

S0

NORMAL SPIRALS

Sa   Sb   Sc

SBa   SBb   SBc

BARRED SPIRALS

The Hubble "Tuning Fork." Hubble arranged the extragalactic nebulae along three intersecting sequences: ellipticals, normal spirals, and barred spirals. Type *S0* was initially hypothetical, but examples were recognized later. (Compare with plate, page 231.)   (*Reproduced from Edwin Hubble:* The Realm of the Nebulae, *New Haven, 1936*)

A barred spiral, N.G.C. 1300. Straight dust lanes emerge from the nucleus and turn abruptly along the spiral arms. The arms in this example form a nearly complete ellipse.   (*Reproduced from Allan Sandage:* The Hubble Atlas of Galaxies, *Carnegie Institution of Washington, 1961*)

the gas being excited to glow by one or more embedded stars; some were predominantly obscure or dark. The dark nebulae were seen as "holes" in the thick background of faint stars, and they could only be seen in projection against star clouds. In some nebulae, bright and dark material were mixed.

Hubble divided the extragalactic nebulae into four major classes: *Irr*, the loose irregular nebulae resembling the Magellanic Clouds; *E*, the elliptical nebulae; *S* and *SB*, the two types of spiral nebulae. With the exception of class *Irr*, the nebulae showed a striking predominance of symmetrical forms.

Some elliptical nebulae were nearly round, others were up to three times longer than wide. Hubble knew that the variety of form among the *E* nebulae might be a combined consequence of variations of their true shapes and their orientations, and he was able to show that the assumption of a single true shape was inconsistent with the observations. Too many appeared round, and he was forced to assume that some of the *E* nebulae were actually spherical.

Aside from the outline shape there was little to distinguish one *E* nebula from another. All nebulae of a given ellipticity appeared to have a similar and very smooth internal distribution of light, and the slight differences that could be noted were very difficult to measure. Therefore Hubble proposed a very simple nomenclature for the ellipticals: the letter "*E*" followed by a single figure numerically equal to ten times the ellipticity, defined as $(a - b)/a$, where a and b are the lengths of the longest and shortest axes, respectively. With this system, round nebulae were designated *E0*; the most highly flattened were *E7*, corresponding to $(a - b)/a = 0.7$ or $b/a = 0.3$, giving a ratio of about three to one between the axes.

In one of the elliptical nebulae Hubble found definite evidence for resolution into stars; their light was cloudy and quite smooth, with one exception. The bright nebula, M87, seemed to be surrounded by a swarm of luminous stars extending well beyond the brighter portions of the nebula itself. Years later, these "stars" were discovered to be globular *clusters* of stars, each individual cluster probably containing hundreds of thousands of stars.

Hubble found a remarkable dichotomy between the elliptical and the spiral nebulae: No ellipticals had an ellipticity greater than 0.7, that is, he did not find any *E8*'s. On the other hand no spirals, when viewed

The Milky Way. This wide-angle photograph covers nearly the entire sky as seen in the Southern Hemisphere. The resemblance to the galaxy shown in the plate on page 236, so striking in this photograph, is difficult to see with the naked eye at night because the eye sees so small a portion of the sky at one time. This photograph was made by A. D. Code and T. E. Houck with the Greenstein-Henyey camera of the Yerkes Observatory. (*Reproduced from* Astrophysical Journal, *Vol. 121* [*1955*], *p. 554*)

from near their edge, appeared to have an actual ellipticity less than 0.7. Thus, spiral patterns appeared to be confined to the most highly flattened of the nebulae, while globular nebulae were invariably smooth and without a trace of spiral pattern.

Classification of the spiral nebulae could not be based on ellipticity because the concept had little meaning for these systems. Many appeared to have the form of a simple disk, but others combined a disk, in which the spiral pattern appeared, with a central bulge resembling an elliptical galaxy. Hubble used the relative size of the nuclear bulge and the spiraling disk as one criterion of classification. Two other criteria were based on the knottiness of the arms, or their apparent resolution into discrete units, and the extent to which the arms appeared to have unwound—or to have wound up—depending on one's point of view.

These criteria varied smoothly from one nebula to another—that is, they were found in all degrees—but in another respect the spirals seemed to show a distinctive separation into two classes. There were "normal" spirals displaying coiled patterns, like cream swirling in a coffee cup; and there were "barred" spirals displaying a bar through the center. Hubble arranged the normal and the barred spirals in two parallel sequences: *Sa, Sb, Sc,* and *SBa, SBb, SBc,* respectively. Often, a spiral arm extended from each end of the bar.

The two sequences of spirals were joined to the sequence of *E* nebulae, giving the appearance of a tuning fork.

In 1936, ten years after his first enunciation of the classification, Hubble published the following descriptions in his book *The Realm of the Nebulae.* They differ very little from the original version:

NORMAL SPIRALS

At the beginning of the sequence (Sa), the normal spiral exhibits a bright, semistellar nucleus (one that appears only slightly less sharp than a star) and a relatively large nuclear region of unresolved nebulosity which closely resembles a lenticular (E7) nebula. The arms which emerge from the periphery are also unresolved and are closely coiled. As the sequence progresses, the arms increase in bulk at the expense of the nuclear region, unwinding as they grow, until in the end they are widely open and the nucleus is inconspicuous. About the middle of the sequence, or slightly earlier, condensations begin to form. The resolution generally appears first in the outer arms and gradually spreads inward until, at the end of the sequence, it reaches the nucleus.

Edge-on view of spiral galaxy N.G.C. 4565. Compare this with the Milky Way as shown in the plate on page 234. The central bulge of this galaxy corresponds to class *Sb* in the Hubble system, and the similarity of the photographs is one reason for supposing the Milky Way to be an *Sb* galaxy. (*Hale Observatories*)

BARRED SPIRALS

The barred spiral is first seen as a lenticular nebula in which the outer regions have condensed into a more or less conspicuous ring of nebulosity concentric with the nucleus, and a broad bar has condensed diametrically across the nucleus from rim to rim. The appearance resembles that of the Greek letter *theta, θ*. As the sequence progresses, the ring appears to break away from the bar at two opposite points, just above the bar at one end and just below the bar at the other . . . and the spiral arms grow out of the free ends of the broken ring. Thereafter the development parallels that of the normal spiral; the arms build up at the expense of the nuclear region, unwinding as they grow; resolution appears first in the outer arms and works inward toward the nucleus.

Hubble suggested calling *Sa*'s the "early" type of spiral and *Sc*'s the "late" type, as a convenience of description, and although he expressly warned against assuming that the sequence was an evolutionary one, Hubble could not deny the possibility. In fact he noted its similarity to a theoretical sequence developed by James Jeans, a British mathematical astronomer. Hubble said: "Although deliberate effort was made to find a descriptive classification which should be entirely independent of theoretical considerations, the results are almost identical with the path of development derived by Jeans from purely theoretical investigations." Actually, Jeans's theory was based on the conception that galaxies would behave like rotating bodies of liquid, and it is now felt that the analogy is too weak to be useful because the individuality of stars has an important influence.

But while he implied that galaxies might be evolving along the sequences, Hubble also mentioned a very different possibility: that they evolve *across* the sequences. If all galaxies first showed bars and then developed spiral patterns, they would evolve from one branch of the tuning fork to another. This question remains unresolved today.

In 1958, Walter Baade, a colleague of Hubble, described his experiences with the classification scheme. While lecturing at the Harvard Observatory, Baade said that Hubble's system "is a very simple one, but . . . there is not much sense in making a system that covers all the little details of spiral structure. . . . I think that the Hubble classification, which deals just with the basic features, conveys all that we want to know." Baade mentioned that he had "searched obstinately" for systems that would not fit the scheme, even among the very faintest galaxies that he could photograph, but the number of systems that really presented difficulties "is so small that I can count it on the fingers of my hand." In one sense, the system was even better than it appeared from Hubble's own work. Hubble classified a number of systems "peculiar" that were later recognized as double galaxies. Baade said: "Hubble had a blind spot where two galaxies were involved instead of one, a case that happened quite frequently. I remember the difficulty that I had in convincing him that a certain galaxy is double. When we finally showed that the two systems had different radial velocities, he still called it a hypothesis. If you eliminate the double systems, I am sure that the number of exceptions is unbelievably small, so efficient is the system."

By 1929, Hubble had estimated the distances of two dozen galaxies, a large enough sample to permit a search for general relationships among them. In that year, he announced a spectacular conclusion: All galaxies were rushing away from each other with a velocity that increases in proportion to the distance between them. Galaxies separated by a distance of 3,000,000 light years recede from each other at three hundred miles per second; galaxies at twice that distance recede twice as rapidly. Evidently the galaxies started with different velocities from a small volume of space, but it is not possible to pinpoint the center of the expansion, because every galaxy would observe precisely the same behavior among its neighbors; every observer could claim that he was located at the center of the expanding universe.

Hubble's work gave sudden relevance to the relativistic speculations of Einstein and others on the nature of the universe; it was possible, for the first time, to plumb the universe to distances beyond the range of Newton's laws of motion. Old papers on relativity took on new interest, and a flood of new papers appeared. What had seemed esoteric speculation became a means of interpreting the most astounding fact of the twentieth century.

Considered in reverse, the expansion of the universe indicated that the galaxies had been very close together five billion years ago. Recent revisions have put the galaxies at somewhat more than twice the distance assigned by Hubble and have extended the time since the start of the expansion to ten or twenty billion years. This estimate agrees well with the estimated ages of the oldest stars in our galaxy and of the solar system. The clear implication from this agreement is that star formation began shortly after the onset of the expansion.

Astronomers are virtually unanimous today that the redshift of the light from galaxies should actually be taken to indicate velocity. Originally, some suggested that the redshift might be due to an in-flight change of the color of individual photons, but the suggestion raised more problems than it answered. Further, the concept of expansion is an integral part of relativity, and this body of concepts has now been verified in several ways.

Astronomers, however, are not unanimous in their interpretation of the expansion, although it appears that critical tests may be performed within the next few years and the issue may be settled. One school contends that all of the matter in the visible universe was condensed in a small volume of space ten or twenty billion years ago and that during the expansion, matter cooled to form stars and galaxies. Whether the expansion will continue indefinitely is not yet clear, although the evidence suggests that the expansion is slowing and will be reversed within the next hundred billion years or so. If this indeed happens, and the universe contracts again into a small volume, we might suppose that it will be reborn when the next expansion commences—like the Egyptian phoenix, a bird that was consumed by fire and then reborn from its own ashes. Philosophically, this is a rather satisfying picture; it reconciles the observed expansion, and the implication that things were drastically different twenty billion years ago, with the notion that the universe must proceed without alteration because it cannot have a limitation in time. If the universe were renewed periodically rather than moving inexorably toward death, atoms would be regenerated and would coalesce again to form new stars. (This idea can be found in Lucretius' poem "The Way Things Are.")

An alternative to this "big bang" theory is the "steady state" theory, which holds that the universe looks the same to all observers and at all times. In the original version of this theory, the outward motion of the galaxies was assumed to be perpetual, and the vacuum was continually replenished by the spontaneous formation of matter, here and there, now and then, at the rate of about one atom per cubic mile per hour.

At first sight, the idea of spontaneous generation of matter appears

239

to be a desperate assumption and one that should be outlawed by the astronomers' rules of conduct. But is it really? Isn't it merely a specific variant of the postulate that matter must have been created *somehow?* The steady-state theory suggests that the creation of matter is continuous; the opponents assume that all matter was created at a distant epoch in the past or that it has always existed—neither of these seems more satisfactory or less satisfactory than the idea of continuous creation.

The contest between the big-bang and the steady-state theories can only be settled by observations of the present universe; if it cannot be settled this way, the distinction will be cast aside as invalid. Although no one yet claims to have proven one theory over the other, there is one phenomenon which may hold the clue to a definitive decision: the so-called "three-degree background radiation" of the universe.

This radiation has the appearance of heat radiation within a furnace set to a temperature of $3°K$ above the absolute of cold. It is much too red to be visible; it can only be detected with extraordinarily sensitive radio receivers, and it comes from the entire sky. This radiation cannot have been produced recently within our own galaxy, and it cannot be explained as a simple accumulation of known radiation from external galaxies because it mimics the radiation of a furnace too well.

The "big-bang" cosmologists were delighted by this discovery because they had a ready explanation: this radiation is the residue of heat from the intense explosion at the creation of the present universe. They said that great quantities of energy, in the form of x-rays and ultraviolet light, were released during the original explosion; as the universe expanded, the temperature of this energy decreased to $3°K$ and the light became redder. (To put it another way, the original radiation was redshifted and cooled by the expansion of the universe.)

The steady-state cosmologists have been ingenious in deriving alternative explanations of the observed radiation, but these alternatives are clearly based on further postulates that would not be needed in the big-bang cosmology. The trouble is that the steady-state theory, in its original form, did not include an explosive event in which the radiation might have been produced, so they have had to cast about for other means of interpreting the observations.

Having discovered the proportionality between distance and recession velocity, Hubble and his colleagues reversed the tables and inferred distances from redshifts. This device has permitted exploration to the edges of the visible universe, and it led to the recognition of the enigmatic "quasars," to which I will return in the Epilogue.

Hubble's last researches comprised a survey of the universe: a census of galaxy types and an attempt to determine the average properties, as well as the extent of deviation, among various types of galaxies. He was also instrumental in the planning for the Hale two-hundred-inch telescope, completed in 1948 and for over twenty years the largest telescope in the world.

Edwin Hubble was a difficult man to get close to because he was a bit of a showman; he enjoyed sucking on his pipe and blowing smoke rings across the table; he spoke with an English accent evidently acquired while studying in Oxford. He knew how to strike a pose. He did all of this with a boldness that won him a few intimate friends, but others were put off by his appearance of self-esteem.

Hubble would have been a great astronomer in any age; that he chose to explore the realm of the nebulae and that he accomplished so much are signs not only of his talent for research but also of his wisdom in choosing a field in which a number of important problems were opening up and in which valuable clues were available.

Hubble's published papers are among the most important in astronomy. They are written with a clear sense of history; they draw on a wide variety of facts; they refrain from speculation beyond a bare minimum. Even after twenty years, they are worth a careful reading.

Hubble died in 1953, and a young astronomer at the Mount Wilson Observatory, Alan Sandage, undertook to collate his manuscripts, notes, and photographs. One result was the *Hubble Atlas of Galaxies,* published in 1961. Sandage's introduction and descriptive material reveal that the essential progress since 1936, when Hubble had published *The Realm of the Nebulae,* was not merely a refinement of classification; it also comprised a wealth of discoveries concerning the structure and content of galaxies.

A complex and suggestive pattern had emerged from the demarcation of bright blue stars *versus* red stars, globular clusters *versus* loose clusters with nebulosity, dust and luminous gas *versus* the general background of stars.

Conceptions of galactic evolution had become more specific—and more varied. Sandage, for example, suggested that evolution might occur *backwards* along the Hubble sequence, from the "late" spirals (*Sc* and *SBc*) through *S0* to the ellipticals. His evidence was the following: Dark material is almost entirely absent from the elliptical nebulae, although it is prominent in normal *Sb* and *Sc* spirals. Dust seems to lie along the inner edge of spiral arms, although it appears on both sides in some cases, giving the appearance of shadows. It is closely associated with bright, blue stars; in fact, bright, blue stars are invariably accompanied by dust and gas.

In Sandage's words, these stars "are known to be very young because their nuclear energy sources can last for only a few million years. Since they are visible today, they must have been created within the last several million years." The *SBb* and *Sb* spirals have bright stars, but they are

> fainter than those in the Irr, Sc, and SBc systems. The presence of dust and highly resolved spiral arms [i.e., many bright stars standing out among the general background of fainter stars] goes hand in hand with other characteristics of the spiral arms. Whenever the arm system is tightly wound, as in Sa and Sb galaxies, there is little or no resolution of the arms into stars, and there is very little dust. Star formation is not going on in these galaxies now; the dust has been used up; and the arms, which were loosely wound and highly branched in the Sc stage, have wound themselves tight against the periphery of the [central] lens [of the galaxy] by the shearing and stretching action of the [rotation of the galaxy].

He goes on to say that the early-type spirals and the ellipticals show virtually no outstandingly bright stars, nor do they show the small emission nebulosities associated with very hot stars. In such galaxies, "star formation has apparently stopped completely, because all the necessary dust has been used up. These galaxies contain stars that are very old."

The implication was that galaxies begin as *Irr, Sc,* or *SBc* systems replete with dust, and with well-opened and highly branched spiral arms; they evolve toward the "earlier" types as the brightest stars die off, the dust is exhausted, and the arms wind more tightly.

Sandage's suggestion that galaxies evolve along the Hubble sequence from "late" to "early" is not universally accepted by astronomers, but the facts on which it was based are not debated. We shall now focus on these.

20 THE CONTENT OF SPIRAL GALAXIES

✲

The first members to be recognized in spiral galaxies were the stars; then followed clouds of gas and dust. Later, the stars were separated into distinct populations, and the peculiarities of the small, central nuclei of galaxies were recognized.

Narrow, dark streaks on the faces of spindly nebulae were, from the first, attributed to dust clouds, and it is now known that many of the less spectacular markings of nebulae and our own Milky Way are also produced by dust. Although astronomers are satisfied that the markings are produced by dust, their proof is based on a network of related arguments, none of which would be convincing by itself.

First, there is the simple fact that many of the markings are very sharply defined, and they *look like* clouds of dust—at least they do now, although they went unrecognized in the Milky Way for centuries because astronomers believed that they were holes in the star clouds.

A more suggestive bit of evidence is found in the bright clouds of gas lying near dark markings. In some nebulosities, a bright central region is surrounded by a dark region, and the simplest explanation is that the larger cloud, composed of dust, has been partially heated and excited by a cluster of hot stars visible in the center. In other cases, a dark region is surrounded by a bright rim of gas and has the appearance of a cloud heated from the outside.

The bitterest pill that astronomers have had to swallow during the twentieth century is the fact that not all of the obscuring matter is collected into obvious clouds. Many galaxies are permeated by an insidious layer of dust concentrated toward the equatorial plane, and the discovery of this layer within the Milky Way was an example of serendipity: an astronomer set out to make a series of measurements using a new technique; he found a peculiar difficulty in accepting the results; he chose to eliminate the difficulty by postulating a layer of dust near the plane of our galaxy, and this postulate then removed a number of other difficulties that had appeared to be unrelated to the first.

Robert J. Trumpler was an astronomer at the Lick Observatory and over a period of years collected all the data available on "galactic" clusters, the loose aggregations of stars which lie quite close to the plane of our galaxy. In particular, he compiled the apparent angular diameters of these clusters and the brightnesses, colors, and spectral types of the cluster members. His aim was to determine the distances of these clusters in order to plot them on a map of the galaxy and compare their distribution with those of the globular clusters and the scattered stars of the Milky Way. He wished, in other words, to repeat the studies of Shapley and others but using a different type of data. In 1930 he published his results, and they were both convincing and distressing.

Cepheid variables had not been discovered in any of the galactic clusters, so Trumpler resorted to another means of distance determination. For each cluster, he made a plot of the apparent brightnesses against the spectral types of the individual stars. (See plate, page 246.) The usefulness of this type of diagram—called the Hertzsprung-Russell diagram—had been suggested two decades earlier in another context by Henry Norris Russell and Ejnar Hertzsprung. They had pointed out

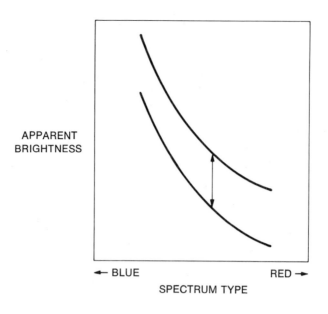

APPARENT
BRIGHTNESS

← BLUE   RED →

SPECTRUM TYPE

The Hertzsprung-Russell diagrams of two galactic clusters at different distances. If the stars of a cluster are arranged by apparent brightness and spectrum type, they will lie along a narrow band. (In some cases, they form a reversed figure "7.") Clusters at different distances will be displaced vertically on such a diagram, because stars of the same absolute brightness will have different apparent brightnesses. The spectrum is not affected by distance, but the vertical displacement between the clusters is a measure of relative distance.

that the stars of the galaxy do not scatter at random on such a diagram, but fill a definite area when true brightness is plotted against spectral type.

Trumpler found that the diagrams of many clusters resembled a reversed figure "7"; a principal sequence sloping up to the left and a few stars standing over on the right. In some clusters, only the principal sequence, or "main sequence," appeared. As the clusters were at different distances from the sun, the H-R diagrams were shifted from one another, corresponding to the greater apparent brightness of stars in nearby clusters. As the spectral types were unaffected by distance, there was no corresponding shift along the other coordinate.

Trumpler could have used the available data in several ways. For example, the *relative* distances of the clusters could have been determined from the shifts in brightness needed to bring the H-R diagrams into coincidence. The absolute distances would then depend on measuring the true distance to at least one cluster. This is the method used most recently by Sandage and others, but Trumpler used a slight variation. He considered the stars as individuals and used a known relationship between true brightness and appearance of the star's spectrum, then finding the ratio of true brightness to apparent brightness. This relationship between true brightness and spectrum is a very subtle one; its discovery was one of the great achievements of the early decades of the twentieth century, because it required the collation of large amounts of data and it has become an extremely valuable tool. At the time of its discovery, no one knew what sort of relationship should exist, although studies of stellar structure had suggested several possibilities and they gave astronomers confidence in using the empirical relationship as a fact of nature. It is now known that the temperature and the pressure of a star's atmosphere are the main determinants of the light it emits; these, in turn, are determined by the mass and the age of the star. In effect, the spectrum is determined solely by the mass and the age of the star.

Trumpler determined the distances of one hundred clusters, and he computed sizes from the angular dimensions of the clusters as seen on photographs. In an attempt to search for possible errors in his work, Trumpler then examined the dependence of the derived size on the distance of the cluster. Expecting to find *no* correlation, he found that, on the average, the nearby clusters were only one-half the size of the distant clusters. The trend from near to far was quite regular; it was not just an illusion caused by one or two extraordinary clusters, nor was there an explanation in any other step of the measurement process. He discarded the possibility that the distant clusters actually *were* larger, because he found a simpler and more natural explanation in interstellar obscuration.

Trumpler assumed that space contains matter which absorbs light. He saw that the absorption of light by matter between the stars, if it occurred, would make the distant clusters appear too faint, and this would give them the appearance of being at too great a distance, thereby accounting for his deriving too great a geometrical size from a given angular size. By trial and error, he concluded that if the dust between the stars were assumed to diminish starlight by a factor one-half for each 3000 light years, he could bring the distant clusters close enough to give them the size of the nearer clusters. For example, a

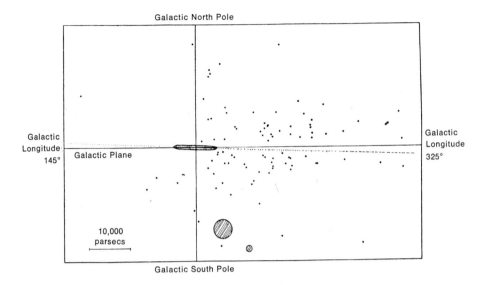

Galactic North Pole

Galactic Longitude 145°

Galactic Longitude 325°

Galactic Plane

10,000 parsecs

Galactic South Pole

The distribution of globular clusters and galactic clusters. In this side view of our galaxy, Trumpler has plotted the globular clusters (*dots*), the galactic clusters (*hatched area*), and the two Magellanic Clouds. It is now surmised that the galactic clusters actually extend to the other side of the galaxy, but that they are obscured by the dust discovered by Trumpler. In much the same way, the distant globular clusters appear to shun the plane of the galaxy, but this appearance is also due to obscuration, which is concentrated near the plane. (*Reproduced from* Lick Observatory Bulletin, *No. 420* [*1930*])

cluster at a distance of 6000 light years—the greatest distance he achieved—would have its light diminished to one-quarter ($\frac{1}{2} \times \frac{1}{2}$), and this diminution would correspond to a factor two in the distance. For a cluster at a very small distance from the sun, the correction would be negligible.

Trumpler also pointed out that the presence of dust among the stars, at least near the plane of the galaxy, might account for the marked redness of distant clusters. Others had suggested that the distant clusters were individually surrounded by clouds of dust, but Trumpler's conception seemed more natural, and it was accepted at once.

Although it was clear, in a qualitative way, that dust among the stars would redden star light, nothing else was known about the dust. The spectroscope was no help, so astronomers had to content themselves with speculation. Many men have had many ideas about the nature of the interstellar dust. Some called it "smoke," and this would seem to be an appropriate name, because some of the material may be the debris of burned-out stars. A respectable fraction of the particles may be magnetic and elongated, because the interstellar clouds polarize the light passing through, and the simplest way to explain the polarization is to assume the presence of a magnetic field. (In fact, such fields have been detected by radio observations.) The detailed nature of the dust grains, and their origin, are still obscure, and every year brings more, rather than fewer, competing solutions to the problem.

Dust may play a crucial role in the formation of stars; the evidence is partly circumstantial: young stars are only found where there is dust. It is also partly theoretical: dust may be more amenable than gas to accumulation into stars, because it can be "blown" by the light of stars into compact clouds which can condense until gravity takes over and pulls it into tight spheres.

Star formation is taking place today in certain galaxies and in certain portions of our own galaxy; dirt heaps may be the birth-sites of successive generations of stars.

A distinctive feature of galaxies is the peculiar distribution of the brightest blue stars and the brightest red stars, and again Trumpler's work on the galactic clusters was an important source of insight. Where star formation appears to be taking place, the brightest stars are blue; in aged, dust-free systems, the brightest stars are red. The distinction is a consequence of the process by which individual stars evolve, and it was clarified in the 1950's and 1960's. Clarification brought restrictions on the assumption of uniformity, a tool that had appeared to work well during the early exploration of extragalactic space.

Trumpler found that, to a degree, bright, blue stars and bright, red stars were mutually exclusive in clusters. The Pleiades cluster, for example, has numerous very bright, blue stars but no bright, red stars; some clusters have the red ones, but the blue ones are not so bright in those cases. Trumpler suggested that the red giants might be the products of old age among the blue giants, but this was pure speculation, and he was not taken seriously because he had no corroborative evidence. Now he is known to have been correct.

Trumpler found that he could arrange the Hertzsprung-Russell diagrams for clusters into a single sequence, and he assumed that the sequence was an evolutionary one. The youngest clusters, according to this picture, displayed a single sequence sloping up to the left, where the blue giants lay. In the older clusters, this band was cut off at the top and the stars were found in the upper right, where the red giants lay. The clear presumption was that the blue giants had been shifted over to the region of the red giants; the older the cluster, the further down on the main sequence would the cut-off take place.

The facts underlying Trumpler's speculation were woven into the fabric of a convincing theory by the discovery of nuclear burning in stars and by extensive calculations after World War II.

This theory lay at the basis of Sandage's contention in the *Hubble Atlas* (preceding chapter) that the "late" spirals with bright, blue giants and much dust should be considered younger than the "early" spirals and the ellipticals without dust or blue giants. Also, this was the theory that seemed to make sense out of Hubble's early discovery that the central, dust-free portion of spiral nebulae seemed to be redder than the spiral arms, in which lay nebulosities and clusters of young blue giants. It was presumed that the central portions of spirals and the elliptical galaxies might be somewhat older than the outer spiraling clouds.

But these facts and theories did not prepare astronomers for a discovery by Walter Baade in the early 1940's: Spiral galaxies contain at least two, and perhaps more, distinct generations of stars. Suddenly, astronomers realized that the principle of uniformity could *not* be applied from the spiral arms to the nucleus; it could not be applied from galactic cluster to globular clusters; in fact, with the passage of time, they saw that it could not even be applied from one cluster to another. Baade's discovery pointed to the need for a new level of sophistication in the interpretation of galactic contents.

By 1936, several puzzling discrepancies had shown up in the investigation of the Andromeda Nebula and other nearby galaxies; certain relationships, expected on the basis of uniformity, had not been verified. The brightest globular clusters of M31 were fainter than they should have been, and a similar difference appeared among the novae. Also, the unusual size of our galaxy remained as an irritation to astronomers; so, although a qualitative match had been found between our own system and that of Andromeda, the numbers didn't seem to match.

The discrepancy concerning the globular clusters was particularly annoying. Even in the presence of obscuring dust, the luminosity of a globular cluster in our galaxy can be determined by comparing it with its variable stars; both will be affected in the same amount, so the relation between the cluster and the stars is unaltered. From the brightness of the variables in M31, the predicted brightness of globular clusters had been computed, but the result was puzzling. The brightest ones were four times fainter than they should have been.

Another fact was puzzling: Hubble had not been able to resolve the central part of M31 into individual stars, although the outer parts had been resolved. Evidently there were none of the very bright, blue stars in the central part. Walter Baade, at the Mount Wilson Observatory, decided to focus on this problem because he felt its solution might also answer some of the other problems.

The prospect of resolution was not very promising, but in the process of examining some unusually sharp photographs, Baade found signs of incipient resolution in an area of amorphous nebulosity. He described the appearance of the region during lectures at Harvard years later: "The plate was very irritating to the eye; a definite structure is emerging all over, but one does not yet see any stars." He was seeing the fluctuations of brightness in a field of stars that were just barely beyond the limits of the one-hundred-inch telescope. As an example of incipient resolution, he cited a diffuse, curved filament that had long been known in M32, the companion of the Andromeda Nebula. This was "weakly visible even under average seeing conditions. On a plate taken under very much better seeing, this feature, normally very soft and diffuse, narrowed up and became much sharper, but was not really different. When resolution was finally achieved, the feature turned out to be one of those accidental chains of stars of nearly equal brightness, with some fainter ones in between. This hazy filament showed up before resolution was finally achieved."

To resolve the central portion of the Andromeda Nebula, it would not suffice to merely lengthen the exposure times, because the brightness of the sky—partly city lights, partly a faint persistent aurora—darkened the plate and set a limit of about ninety minutes for the films he was using at the time. They were films particularly sensitive to blue light; Baade guessed that if he shifted to the red-sensitive plates and used colored filters to eliminate the auroral light, he might be able to catch the stars of the Andromeda Nebula—if he could expose for nine hours!

The slightest slip in the focus of these plates would blur the light of the stars and blend it into the general illumination of the background and the aurora. He computed that he would have to hold the focus of the one-hundred-inch telescope to 0.1 mm, this despite the fact that the focus might change by as much as several millimeters on nights of drastic temperature change. So he tried to swing the plate out of the way and measure the focus every hour or so. But each measurement took half an hour, so he decided it was much too inefficient. Then he tried an alternative.

During a long exposure, the astronomer guides the telescope by watching the image of a star through a microscope equipped with a pair of cross-hairs. If the star wanders, the telescope is brought back into position with a tiny motor, or the film is moved to compensate.

Because Baade's guide star was not exactly at the center of the field of view—where the photographic plate was set—its image was slightly distorted. This is inevitable for a telescope with a parabolic mirror, and it is known as the "coma" of the telescope, because the distorted image looks somewhat like a comet with a bright nucleus and a fan-shaped tail. From the shape of the image, Baade determined whether the photographic plate needed to be moved inward or outward to sharpen the image, and how much.

He recorded:

> I remember the first night very well; I selected intentionally a night when the seeing would presumably not be first-rate. I still remember how confused I was as to what to do. But you just calm down and wait for a moment of good seeing; and after you have mastered it, it is astonishing how little time you need to see what the situation is, and to give a quick turn to the focus. . . . The whole trick is in dealing with the occasional disturbance. I can imagine that some people would lose their heads and get nervous and angry in a short time, and would begin to turn the focus violently.

Another problem was to catch telescope mirrors in their proper shape. Two mirrors were involved: the one-hundred-inch "primary" mirror, whose parabolic shape focused the light; and a smaller, flat mirror, named the "Newtonian secondary" after its inventor. The

Newtonian secondary deflected the light through the side of the telescope to the photographic plate. The secondary, being higher in the
dome, is more susceptible to heat during the daytime hours. Baade says:
"Since we have no insulation, the only thing to do was to open the dome
early—right after lunch time—and turn it around so that no sunlight
could fall directly on the instrument. . . . Usually in the time between
1 P.M. and 10 P.M. [when the instrument was to be used] the secondary
mirror had a chance to get into equilibrium again."

In his Harvard lectures, Baade said the preparation for the attack
on the Andromeda Nebula lasted from the fall of 1942 to the fall of
1943. Those were dark nights on Mount Wilson because the Los
Angeles valley was in blackout. Baade said: "After taking all these
precautions, the resolution was a very simple matter. In August and
September I resolved the central part of M31 and the two nearby
companions in rapid succession. . . . After the shooting was over, it
was quite clear that all the precautions had actually been necessary; I
had just managed to get in under the wire, with nothing to spare."

To Baade, the most remarkable aspect of the whole affair was that
"stars appear in very large numbers—thousands and tens of thousands," once the resolution had been achieved. The elliptical companions and the central regions of M31 comprised great numbers of red
stars; they were much brighter than the red stars found by Trumpler
among the galactic clusters, and for this reason Baade's gamble had
paid off. (See plate, page 254.)

It was obvious at once that the Hertzsprung-Russell diagram for
these stars was entirely different from those found by Trumpler. To
what could it be compared?

Baade tells: "I had always been very much interested in the
globular clusters, but in those days I had almost forgotten them. After a
long time it finally dawned on me that there is another concentration
[of stars] in the H-R diagram, what we now call the giant branch of the
globular clusters." He realized that the bright-red stars of the resolved
nebula were similar to the stars of globular clusters. "The moment I
thought of that, everything began to make sense." (See plate, page
256.)

The discovery of these bright, red stars was not sufficient to clinch
the identification of the contents of these nebulae with the stars of

globular clusters; the next step was to look for variables like those in globular clusters: the Cepheids of a period of less than one day. But these stars were only one-tenth as bright as the stars Baade had barely managed to catch, so they were clearly out of reach. The clincher came from an unexpected quarter.

Several years earlier, Shapley had announced the discovery of a new type of stellar system, exemplified by two sparse clusters of stars known as the Sculptor and Fornax systems. They were outside our galaxy, and yet they resembled globular clusters; that is, they had abundant red giants, and they also contained the short-period Cepheids. They were immense compared to ordinary globular clusters, yet they were dwarfish compared to elliptical galaxies.

Hubble and Baade had studied these systems, but at the time they did not see them for what they were: the "missing link" between globular clusters and elliptical galaxies. They saw them merely as globular clusters. Baade says: "nobody had the wits at the time to see that the whole problem was solved and we knew what the E galaxies consisted of. This strenuous detour from another angle [into the Andromeda Nebula] was needed before we saw the light."

Thus, with the recognition that the globular clusters, the Sculptor and Fornax systems, the central portions of the Andromeda Nebula, and the E galaxies were all similar in their stellar population, the key had been found to removing the discrepancies. Baade designated these the second population and their members were called "type II" stars.

In contrast with these stood the stars of the disks of spiral galaxies; these became "type I" stars, and it is to this population that the sun belongs.

The recognition of this dichotomy was soon followed by a sequence of other discoveries; distinctions that had been invisible became

Resolution of an elliptical galaxy, N.G.C. 205. This galaxy is the larger of the two companions of the Andromeda Nebula (plate, page 55), where N.G.C. 205 appears to the upper left. The brightest stars of this galaxy are all of nearly the same brightness. In the early 1940's, Walter Baade managed to resolve this galaxy and the central region of the Andromeda Nebula into stars. This accomplishment revealed that galaxies contain at least two distinct generations of stars. (*Hale Observatories*)

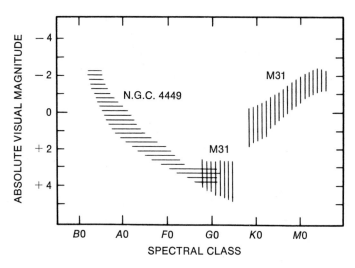

Hertzsprung-Russell diagrams for the two populations. In the central regions of M31, the Andromeda Nebula, the brightest stars are of classes K and M, corresponding to low temperature and red color. Baade called this the type II population of stars to distinguish it from a population like that of N.G.C. 4449, in which the brightest stars are of types B and A, corresponding to high temperature and blue color. This dichotomy is understood by supposing that the type II stars are much older than the type I stars. (*Reproduced from O. Struve and V. Zebergs:* Astronomy of the Twentieth Century, *New York, 1962*)

obvious. One of these was distressing: the Cepheid variables did *not* all obey the same period-luminosity relation.

The assumption of uniformity had been a false hope, and it had led to many of the discrepancies. Cepheids of globular clusters (type II stars) were four times fainter than Cepheids in the disk of our galaxy (type I). Distances based on the type I Cepheids had to be increased by a factor two; the Andromeda Nebula was doubled in the eyes of astronomers, and it became comparable to our own system at last. The distances to fainter galaxies were increased by a similar ratio and the age of the universe, as computed from its expansion, was doubled—giving plenty of time for the comfort of geologists.

It is a simplification to say that there are *two* types of stellar population, because some astronomers hold that there is a variety of intermediate types as well. Why is there not just one type?

Evidently, the first stars created were of type II; they are known to be poor in elements heavier than hydrogen and helium, the lightest elements and simplest atoms. Astronomers conjecture—and they can adduce an abundance of corroborative evidence—that these first generation stars produced heavier elements and then spewed them back into space, where the replenished gas and dust formed a second generation, the population I stars. Thus, the difference between the H-R diagrams (the absence of blue giants in one and of red giants in the other) is attributed to difference of chemical composition and age, as is the difference by a factor four in the brightness of the Cepheid variables.

Much is still unknown: how the elements are returned to the interstellar space and collected into new stars; why the population I

The spiral galaxy M104. The central region of this galaxy comprises stars of type II, while the disk comprises stars of type I. On a photograph such as this, there is no doubt that the dark peripheral lane of dust lies on the near side of the galaxy. (*Hale Observatories*)

stars (the younger ones) are confined to the disk of galaxies, while the population II stars appear in more nearly spherical arrays. Some think that our galaxy was originally more spherical; the older population was formed in this shape, and the dust and gas then flattened, before forming the more recent population.

All galactic types contain very old stars. Even the late spirals (*Sc* and *SBc*), whose arms consist of condensations of young stars, contain a small nucleus of stars as old as those in elliptical galaxies—or so it appears. The implication is that star formation started at the same early epoch in all of these galaxies, and that, although it has ceased in the *E*'s, it is still underway in the spirals.

Evidently something is happening in the spirals to maintain the birth of stars. Do the arms themselves enhance the formation of dust and hence the birth of stars? Or is this looking at the problem from the wrong end? Perhaps the presence of dust leads to the formation of spiral arms. Is rapid rotation enough to cause the formation of spiral arms?

If so, what about the *S0* galaxies, in which there is no spiral pattern but in which the presence of a flattened disk indicates rapid rotation? Hubble introduced this class as a transition between the *E*'s and the spirals. An *S0* galaxy has the same brightness distribution as an ordinary spiral galaxy with the arms removed; that is, a nucleus, an elliptical central lens, and a highly flattened disk. In some *S0* galaxies, a few weak dust lanes can be seen, but the class as a whole is defined by the absence of spiral arms. Baade noted that this type is commonly found in clusters of galaxies, and Lyman Spitzer, of Princeton, suggested that they might be "stripped" galaxies, in which the dust and gas required for the formation of spiral patterns had been removed by collisions among the galaxies.

Clustering is common among galaxies; they appear in pairs, in triplets, etc., and in vast assemblies containing hundreds and thousands. Our own galaxy belongs to a group of nearly twenty recognized members. They are fairly well separated from each other, although the Andromeda Nebula has two small companions, and our galaxy is accompanied by the two Magellanic Clouds. In some clusters, the members are more closely packed and they must collide from time to time.

What happens when two galaxies collide? Speculation cannot take us very far, but one point is clear. If the colliding galaxies consisted only of stars, a collision would be a very quiet affair, like the mutual penetration of two ghosts. A galaxy of stars is simply empty space with points of light, and the points would almost certainly pass by each other during a "collision."

But this would not be true for galaxies containing dust and gas. When matter is finely divided it has more area, and two clouds of dust could not pass through each other without collision. If a magnetic field were present, as seems very likely in galaxies containing gas, the motion of one galaxy would produce electrical currents in the other, and the collision might be a spectacular affair.

But even if the dust and gas interacted violently during a collision, the stars would continue on their original paths, leaving the smaller stuff behind in a chaotic cloud to mark the point of collision. This, at least, was the hypothesis put forth by Baade and Spitzer during the 1940's.

But to say that S0 galaxies represent spiral galaxies which have been stripped, and in which the arms have faded, presents two difficulties. In the first place, such galaxies are also found outside of clusters, in regions where they could not have recently undergone a collision. In the second place, a recent theory suggests that spiral patterns can arise and persist in the absence of dust; if this theory is correct, then the mere absence of dust may not be sufficient to explain S0 galaxies.

Up to now, I have spoken as though the apparent flattening of a galaxy necessarily implied rapid rotation. This is true, but it must not be inferred that the galaxy behaves like a drop of liquid which is flattened by the centrifugal force of rotation. A galaxy is a swarm of stars, and the analogy to a liquid is a false one; we must look a bit more closely at the meaning of the word "rotation."

The stars in a flattened system, like the planets of the solar system, all pursue orbits lying nearly in the same plane; this much is clear. But what of the globular swarm? There, the individual stars pursue orbits which, individually, resemble the orbits within a flattened system, but they are arranged in a random pattern rather than lying nearly in the same plane. The sphericity arises from the chaotic pattern of orientation. The stars themselves may move just as rapidly in one system as the other, but in a flattened system they all move together as though they were parading.

Aside from its effect on the external form of a galaxy, this organization of the motions also has a direct bearing on the internal structure. The process is a subtle one, and astronomers have not yet come to a consensus concerning the details of its operation, but two points of view have been expressed, each emphasizing a different aspect of the dynamical state within a galaxy. It is possible that both are partially correct. They are the following:

Stars moving in random directions within a spherical galaxy will not maintain fixed positions relative to each other. Even if there were a tendency for an internal pattern to arise, the pattern could not intensify, because the only coherence among these stars is the tendency to be attracted gravitationally toward the common center. This tendency influences the external form, but it cannot produce and maintain relationships among the stars *within* the galaxy. Thus, spherical galaxies should not be expected to display spiral patterns—as long as the assumption of internal chaos is valid.

Within a flattened galaxy, however, the stars at a particular point are moving in very nearly the same direction—in circles about the center. Stars that are born together will remain together for a while, although the stars nearer to the center tend to complete their orbit in shorter times than the outer stars.

Near the center of some galaxies, the rotation has the appearance of a solid wheel, all stars requiring precisely the same time to make one revolution. The general tendency, however, is for the outer stars to require longer.

Suppose a large cluster of stars is formed, either by the clustering of existing stars or by the multiple birth of many stars in one neighborhood. As the cluster moves about the center of the galaxy, its inner portions will move faster than the outer portions, and the cluster will be drawn out into an elongated pattern which will resemble a spiral arm.

The 200-inch Hale telescope on Mount Palomar, California. This is the last of the series of great telescopes whose conceptions and construction were inspired by George Ellery Hale. (*Hale Observatories*)

As it ages, the "arm" will become longer or more tightly wound about the center of the galaxy. Finally, the condensation will vanish into the general background of stars. This is one theory of the formation of arms: condensations are drawn out by the rotation of the galaxy.

Another theory, originally espoused by the Swedish astronomer Bertil Lindblad, was developed in a modified form by C. C. Lin of M.I.T. and his student, F. Shu. This theory holds that the arms are not clumps that move along with the stars, but are regions through which the stars pass, and in which they congregate, rather like autos at the intersection of two roads.

The density of autos is usually higher at an intersection than on the open road. Because cars tend to interfere with each other, drivers decelerate at an intersection, even in the absence of signals, because they know they must take turns passing through. On the open road, such condensations of traffic can occur in the wake of an accident. Traffic temporarily halts; cars line up until the wreckage has been removed; then, the lead car starts up, but it may be an hour before the condensation has totally dissolved.

According to the Lin-Shu theory, stars tend to bunch within a galactic traffic pattern; the consequent deceleration leads to the appearance of spiral patterns. This is a collective behavior and, although the stars influence each other by gravitational interaction rather than anxiety, the effect is quite similar to bunching in traffic.

Lindblad claimed that the spirals are wound in a sense opposite to the one which most astronomers had intuitively guessed; he said that the arms lead in the rotation rather than trail. Lin and Shu, on the other hand, say that the arms will trail, but that the pattern is self-sustaining and its properties change very slowly. The specific shape will depend, according to their theory, on the flattening and the density of the galaxy.

Self-sustaining patterns are not uncommon on the earth. The growth of cities is one example: people gather together for a variety of reasons and they become dependent on each other. Many traditions and cultural habits may be viewed as self-sustaining patterns; once started, these habits tend to perpetuate themselves.

Other types of theory have also been developed, but for the most part they have started from a more elaborate set of assumptions and

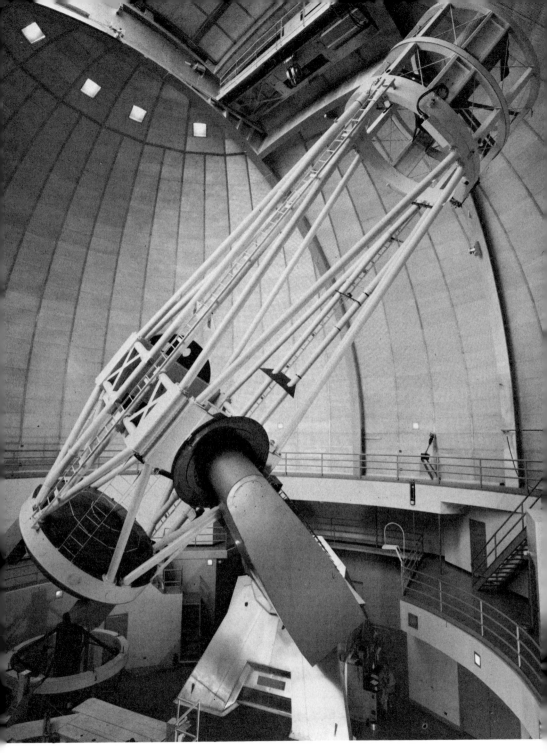

The 120-inch telescope of the Lick Observatory on Mount Hamilton, California. (*Lick Observatory, University of California*)

have been more difficult to test with calculations. According to one, the gas and dust of a galaxy are controlled by a rotating magnetic field, and the dust produces stars. At present, it appears that the known magnetic fields are too weak to confine the material into arms. In another hypothesis, matter is ejected from the center of the galaxy, as though it were shot from a nozzle, and the rotation of the nozzle leads to arm formation. A theory such as this may appear bizarre, but it is no more bizarre than the facts: jets of gas have been observed streaming from the center of our own galaxy.

Such jets may have nothing to do with the form of our galaxy, but their presence suggests that no hypothesis can be rejected merely because it sounds incredible. We will need some incredible hypotheses before the facts will be collated into a credible scheme.

21 THE SPIRAL FORM OF THE MILKY WAY

It was not by sudden insight that astronomers came to recognize the Milky Way as a spiral galaxy; it was through an accumulation of guesses and corroborations. In 1919, Shapley claimed that the globular star clusters lay in a great cloud about the center of our galaxy, but he called the Milky Way a metagalaxy and envisioned it as a swarm of lesser systems. In 1925, Hubble identified the spiral nebulae as island universes. In 1927, the rotation of the Milky Way about the star clouds of the constellation Sagittarius was demonstrated.

By 1940, few astronomers doubted that ours was a spiral galaxy, but none could point to an example and say: "That's what ours is like." The favorite candidates for comparison with our own were the Andromeda Nebula (classed *Sb* by Hubble) and M101 (of class *Sc*). The second has a small nucleus and loose arms; the first has a larger nucleus and tightly wound arms. Because our perspective is so hindered by our location, astronomers are still uncertain of the precise shape of our galaxy, but examination of extragalactic nebulae provided clues to the spiral arms within our own galaxy.

Nebulosities, bright blue stars, clusters, Cepheid variables of long period were found to lie along the arms of the Andromeda Nebula and similar spirals. Astronomers sought them within our Milky Way, and they soon discovered indications of spiral arms in the neighborhood of the sun. But the picture was fragmentary. Then, during World War II a young Dutch astronomer, H. C. van de Hulst, suggested that hydrogen atoms in space—if there were any—might emit radio waves which could be detected on earth. In 1951, the signals were detected in America and in Holland; they consisted of a pure note like the sound of a tuning fork, whose Doppler shift would reveal the velocity of the emitting gas. The fact that hydrogen emitted a single frequency of radio radiation made these signals extraordinarily valuable, because the distance of the emitting clouds could be determined from their velocities—on the plausible assumption that the hydrogen gas conformed to the smooth rotation of the galaxy.

Suddenly the map of the Milky Way was filled with details; some of them coincided nicely with the previous indications of spiral arms. Discrepancies were found, but they were not too disconcerting; they revealed that a spiral arm is a complex arrangement of stars, gas, dust, and clusters, not all of which are co-extensive.

The final step was Baade's demarcation of the two extremes of stellar population (previous chapter), which provided a scheme for unraveling the patterns of distribution that had been found among our neighboring stars: very young stars were confined to spiral arms, old stars were not; the sun moves in a circular orbit about the center of the galaxy, other stars, some of them very poor in metallic content, move in highly elliptical orbits that take them much closer to the center. Astronomers began to see these variations as indications of the past history of our galaxy.

To me, the single most important conception underlying these discoveries is the rotation of the galaxy about a definite center. Each search for data with which to classify our galaxy in the Hubble scheme depended in one way or another on the nature of this rotation. Its discovery came about in this way.

.   .   .

In the early decades of the twentieth century the most widely discussed enigmas were concerned with the motions of stars through space. Two peculiarities were noted, and they had no apparent correlation. A small number of stars move at an extremely high velocity, up to fifty miles per second, through the neighborhood of the sun—like hawks flying through a flock of doves. These were the "high-velocity" stars, and their origin was totally unknown. In addition to the peculiar motion of this small fraction of the stellar population, there was a tendency for all the stars of the solar neighborhood to divide themselves into two opposing streams, moving at about one-fifth the speed of the high-velocity stars. One of the streams moved toward the region of the star clouds in Sagittarius, the other moved away from that region into a rather sparse portion of the Milky Way. (See plate, page 268.) Some astronomers took this dichotomy as proof of the ring structure of the Milky Way; they said we were observing the stars moving in both directions about the ring.

In one of the finest strokes of astronomical insight of the twentieth century, the Swedish astronomer Lindblad proposed another solution to the problem. In 1921, he showed that these two peculiarities were different aspects of the same fact: The Milky Way is in rotation about a point in the direction of Sagittarius.

Lindblad proposed that our galaxy comprised a number of subsystems, each of a different shape and each rotating with a different velocity about a single galactic center. The globular clusters, for example, belonged to a spherical system which rotated very slowly, while the stars of the flat disk of our galaxy comprised a rapidly rotating system. According to this model, the so-called high-velocity stars were actually moving at a small velocity about the galaxy; they appeared to move rapidly because we were overtaking them.

Why do they move slowly? Lindblad pointed out that the slow-moving stars must be in elliptical orbits; their low speed implies that these stars are being drawn toward the center of the galaxy rather than continuing in a circular orbit at a constant distance from the center of the galaxy. (In a similar fashion if the earth were slowed in its orbit about the sun, it would be carried toward the sun and its orbit would become elliptical, with the outer extremity touching the present orbit.)

Lindblad further suggested that the two star-streams might merely result from the inward and outward motions of stars following an array of such elliptical motions. At any one location, roughly half of the stars would be on the outward-moving portion of their ellipse; the other half would be moving inward. The net effect, seen from the sun moving along with the crowd (so that the greatest part of the orbital motion is undetected), would be two star-streams, one moving toward the center of rotation and the other moving away. By contrast, the high-velocity stars would appear to move backwards, at right angles to the direction of the galactic center.

All of this was hypothetical, although Lindblad provided a mathematical analysis to support his contention. Then in 1927, Jan Oort, a Dutch astronomer, confirmed Lindblad's hypothesis and showed that most of the sun's neighbors move in the manner of planets about a

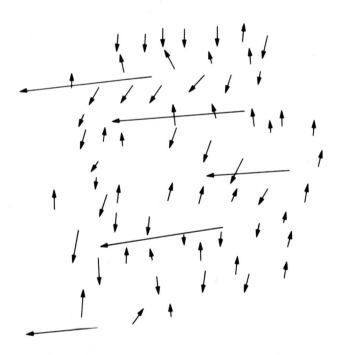

Motions of stars near the sun. At the start of the twentieth century, stars seemed to be divisible into three groups, illustrated here. Two groups (the two "star streams") moved slowly and in opposite directions; a third group (the "high-velocity" stars) moved much more swiftly and at right angles to the other groups. During the 1920's the cause of this pattern was traced to the rotation of our galaxy about a point in the direction of Sagittarius.

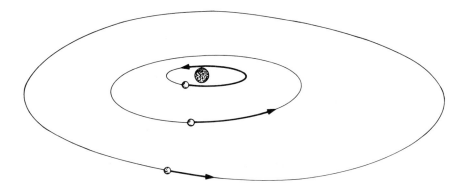

Planetary orbits. The outer planets move more slowly than the inner planets because they are farther from the central mass and feel a weaker gravitational attraction. By analogy, the outer regions of our galaxy move more slowly than the region near the sun. However, the analogy does not hold all the way to the center of our galaxy, because the mass of our galaxy is carried in the stars and they are not highly concentrated to the center.

central sun. Oort did not suggest that there actually was a massive nucleus in the center of the galaxy pulling the stars into orbit, the way the sun pulls the planets. He realized that the accumulated attraction of all stars with orbits smaller than the sun's would be similar to that of a central nucleus. Oort's analysis provided observational proof of Lindblad's hypothesis; to understand how, we may start by considering the motions within the solar system.

Because the sun is far more massive than the planets, it remains fixed at the center of the solar system, and the orbits of the planets are concentric and focused on the sun. (See plate above.) The closest planet, Mercury, is attracted most strongly, and to remain in a nearly circular orbit it must move most swiftly. Kepler discovered the laws describing the motions, and one of these laws states that the square of the planetary period varies as the cube of the radius of the corresponding orbit. (That is, P^2 varies as r^3, where P is the period and r is the radius of the orbit, or one-half the major axis in the case of an elliptical orbit.)

The period of Mercury's orbit about the sun is eighty-eight days, or 0.24 years; from this we may compute the size of Mercury's orbit. We have, from Kepler's law, $(0.24)^2 = r^3$, so the numerical value of r is 0.39 of the earth's orbital radius; hence $r = 0.39 \times 93$ million $= 36$ million miles.

From geometry alone, we know that the velocity of a planet varies as r/P, and we conclude that the velocity decreases continually with increasing distance from the center, as $1/r^{\frac{1}{2}}$.

The situation is a little different in a galaxy, because the mass is not all concentrated in a central body; it is distributed among the stars. Consequently the interior stars do not feel as large an attraction as they would if the mass were concentrated; they tend to move more slowly in their orbits.

But the outermost stars behave as though the entire mass were in a single body, and Oort based his analysis on the assumption that the sun was in the outer reaches of our galaxy. If this were indeed the case, the stars beyond us would move more slowly while those nearer the center of the galaxy would move more rapidly. Examining spectroscopic evidence for radial velocities among stars, Oort found this pattern confirmed in a remarkably definite fashion. Stars directly ahead of us and directly behind us—that is, in the same orbit—moved with very nearly our velocity, and they appeared to be stationary with respect to the sun; they displayed a zero radial velocity. Also, stars lying directly between us and the center of our galaxy, as well as those in the opposite direction, appeared to be moving directly across the line of sight; they too displayed a zero radial velocity. Stars in intermediate directions either receded or approached, depending on the particular direction. (See plate, page 271.)

This pattern implied rotation about a point lying in either of two directions: toward the system of globular clusters studied by Shapley, or in the opposite direction. Oort assumed that the center lay in the direction of the clusters, and he was able to compute its distance from the detailed manner in which the stellar velocities varied near the sun.

His result, 20,000 light years, was only one-third the figure advocated by Shapley. The discrepancy was clearly serious, but nothing could be done to clarify it at the time; Oort's work seemed to be evidence in favor of the Shapley universe, but the discrepancy of scale was disconcerting. Three years later, when Trumpler showed that the distant clusters were affected by obscuration and looked farther than they were, the discrepancy vanished.

Thus Oort confirmed Lindblad's hypothesis of a large-scale rotation within our galaxy. His analysis went further; he identified several of the subsystems postulated by Lindblad, showing that the stars of the

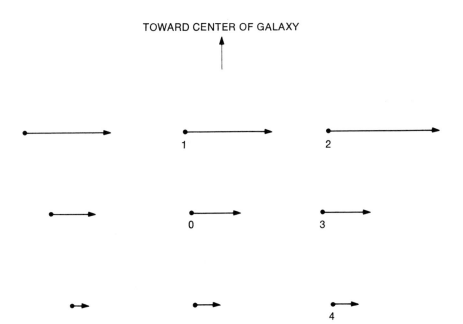

TOWARD CENTER OF GALAXY

Schematic representation of the galactic rotation near the sun. Stars nearer to the center of the galaxy move more swiftly than the sun; a star at point 2 will recede from the sun. A star at point 3 moves with the sun's velocity. Stars at points 1 and 5 move across the line of sight and have a vanishing radial velocity.

spherical system differed from those of the highly flattened system. Stellar properties appeared to be correlated with location and motion through the galaxy.

Early in these studies, it was even suggested that velocity itself could alter the appearance of a star, but the differences were soon attributed to variations of chemical composition, and astronomers realized that the composition of a star depends on the time and place of birth, as well as on the time elapsed since the birth.

Twenty years later, when Baade announced his discovery that stars of elliptical galaxies, the globular clusters, and the central regions of spiral galaxies, should be considered as a population distinct from stars of the disk and arms of a spiral galaxy, he pointed out that this dichotomy was precisely the one proposed earlier by Oort for our own galaxy. The high-velocity stars corresponded to Baade's type II, while the sun and its neighbors were representative of type I. The consensus is that the two populations were born at different times. If type II stars were formed while the galaxy still retained a large spherical halo of gas and dust, this would account for the present spherical distribution of, say, the globular clusters, which are the finest examples of type II. It would also account for the low abundance of the heavier elements in these stars, elements that evidently became more abundant later in the life of our galaxy. The type I stars, which are confined to the disk of a spiral galaxy, were evidently formed later, when the halo had collapsed and the dust within the galaxy had condensed to a flattened system. Stars born in this flat system would remain near the disk.

Intermediate types have also been discovered, so the dichotomy is not complete. But it is sufficiently marked to prove that star formation may span an important fraction of the age of a galaxy.

Lindblad, the man who had resolved the enigmas of the high-velocity stars and the two star-streams, conceived a theory of the formation and development of spiral arms. This theory suggested that the spiral galaxies should be rotating with the arms *leading;* and according to his theory, because the central parts rotate more quickly, the arms should be unwinding. This prediction was contradicted by V. M. Slipher, an American astronomer. He had examined the spirals with the spectrograph to determine which side was approaching the earth. He found, or thought he found, that the spirals were rotating with the arms *trailing.*

In the 1930's, Hubble examined virtually every bright spiral visible from Mount Wilson; one purpose was to determine the sense of rotation—if indeed the spirals did all rotate in the same sense. He remarked in 1943 that the "empirical solution of this apparently simple problem is curiously difficult," and he divided it into three separate problems.

First, the spectrograph could distinguish the approaching side from the receding side if the plane of the galaxy in question were not tilted much more than 60° from the line of sight. Second, an examination of the spiral pattern would distinguish the right-handed spirals from the left-handed spirals. (On the photograph of a right-handed spiral, a point moving inward along the spiral pattern would move clockwise.) The distinction between right- and lefthandedness was very difficult for spirals in which the plane was tilted less than 10° or 20°.

The hooker came with the third step: determining which side of the spiral was closer to the earth. As an examination of the plate

Schematic diagram of a spiral galaxy. According to Hubble's definition, this is a right-handed spiral because a point moving inward along an arm moves clockwise. The spectroscope can tell which end is approaching the earth. Suppose it is the right end. To decide whether the spiral arms are trailing or leading the rotation, we must discern whether the lower side is the closer or the farther side. Hubble's criteria for distinguishing the near side were not universally accepted, but his conclusions were verified when the spiral pattern of our own galaxy was discovered.

above indicates, a right-handed spiral in which the right end of the major axis approaches the earth could have its arms either trailing or leading. Its arms trail if the upper end of the short axis is the closer; they lead in the contrary case. Debate focused on techniques for deciding which was the closer side of the spirals, because in all but a few cases those spirals which were tilted enough to reveal their patterns were tilted too much to permit an unambiguous identification of the closer side.

The star clouds and gas clouds within a galaxy do not obscure each other or interfere with each other's light. Thus, if we imagine gas to be distributed through *one side* of a star cloud, we cannot tell from appearance alone whether the gas is in the near side or the far side. Dust behaves differently; a dust cloud on the near side of a star will obscure the star's light, while it cannot affect the light if it lies behind the star.

This unique property of dust made it the astronomers' only hope for identifying the near side of galaxies. Dust lanes are a common feature of spiral galaxies, and when they cut across the central portion of a galaxy, they give a clear indication of the near side. Unfortunately, the spiral pattern is rarely invisible in such a galaxy, so it is no help to know which side is closer. The trick is to find a galaxy tilted just sufficiently to reveal its spiral and at the same time not tilted too much to permit determination of the near side. (See plate, page 274.)

With a slight tilt, the sharply marked dust lane moves off the central lens, leaving a residual obscuration which appears to predomi-

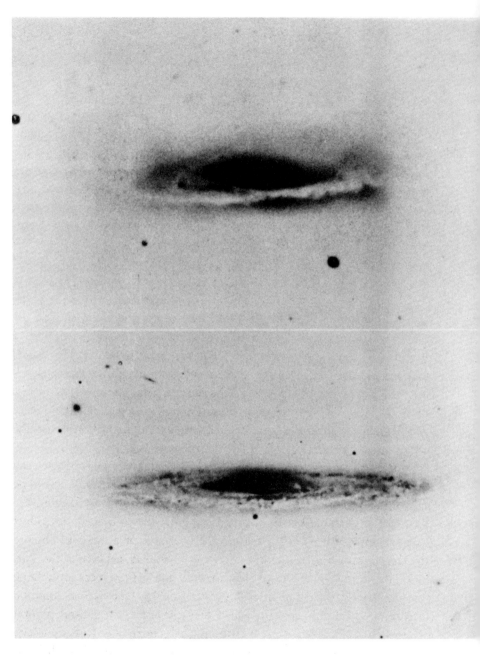

Negative photographs of two spiral galaxies tilted far enough so that the spiral pattern can be distinguished and yet tilted little enough so that the near side can be unambiguously distinguished (according to Hubble). The upper galaxy, N.G.C. 3190, shows an extraordinary twist which aids the determination of the spiral pattern. This was the galaxy which seems to have convinced Hubble that the arms trail. (*Reproduced from the* Astrophysical Journal, *Vol. 97* [*1952*], *p. 114*)

nate on one side. In the course of his survey Hubble had been able to show that, depending on whether it is assumed that the obscured side is the near side or the far side, the arms either trail in all spirals or lead in all spirals.

Hubble claimed that the obscured side is the far side; Lindblad claimed the opposite. Close examination of their papers reveals that the two men were in agreement concerning the empirical facts but differed in a crucial assumption; Hubble assumed that the dust was confined to a narrow sheet near the central plane of a galaxy; Lindblad held that it was distributed in a cloud that enveloped the entire galaxy and extended beyond the limits of the central bulge. Further, Lindblad assumed that the dust was thoroughly mixed with the stars in a spiral arm, while Hubble saw the dust as being confined to the inner edges of the arms.

Hubble's arguments appear to have persuaded most astronomers that the spiral arms trail. (I was a student during the height of the debate, and I remember feeling that Hubble was clearly correct. At the time, Lindblad's interpretation seemed somewhat artificial, and I did not appreciate the force of his theoretical arguments.)

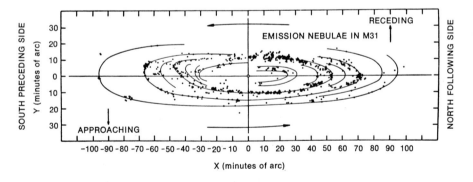

Tracing the spiral pattern of the Andromeda Nebula, M31. On this diagram, the locations of emission nebulae within the Andromeda Nebula are plotted. The sense of rotation, as determined by the spectrograph, is also indicated. On this plot, it is difficult to tell whether the spiral should be considered left-handed or right-handed (plate, page 273), but when the points are replotted as though we were viewing the galaxy from above, the pattern is clearly left-handed. Thus, M31 rotates with its arms trailing. (*Diagram by Walter Baade and Halton Arp, reproduced from the* Astrophysical Journal, *Vol. 139 [1964], p. 1029*)

A decision came when the arms were traced within our own galaxy: they trailed. Hubble had been correct in the case of our galaxy; and because he had shown that the arms of all nebulae wind in the same sense, it was recognized that they trail in all spirals. This was one of the few cases in which astronomers have interpreted the structures of other galaxies by examining our own. (See plate, page 277.)

Visual observations of spiral nebulae often indicate a starlike nucleus. Recently, it has become clear that the nuclei are not only the geometrical centers of galaxies, they may also play a central role in the birth and development of galaxies.

During the 1940's, the Soviet astronomer V. A. Ambartsumian suggested that enormous quantities of matter, for the most part invisible, are being ejected from the centers of many galaxies. His suggestion seemed incredible at the time, but it was based on some undeniable facts. He did not claim to explain *why* the matter was being ejected; he put forth the speculation as a framework for further research. In a sense the idea was beyond arguing, because the evidence one way or another was too slim, but many astronomers have kept it in the back of their minds and they now think that it is being confirmed.

With a very short exposure in a large telescope, the Andromeda Nebula resembles a small oval hardly larger than a star. This is the image of its nucleus; its light is very intense. Its oval shape indicates that it is rotating rapidly, and its color is rather similar to the color of the population II stars in the much larger central bulge of this galaxy.

The nucleus of our own galaxy is invisible on ordinary photographs because the dust clouds of the Milky Way hide it from us. A search of the region of the globular clusters was made with radiation that can reach us through the dust. A small bright area was discovered at the expected location; it is about one-twentieth of a degree across and this is exactly the size that the Andromeda Nebula's nucleus would appear at an equal distance.

The radiation from the nucleus of our galaxy is not simply star light, however—its radio strength eliminates that possibility. Evidently the center of our galaxy is a generator of "synchrotron" radiation, the type of radiation emitted by electrons spinning within the magnetic field of a synchrotron, a form of atomic accelerator. If an electron is shot into a magnetic field, it will spiral. A cloud of spiraling electrons will emit radiation at all wavelengths, from the visible to the radio, and the

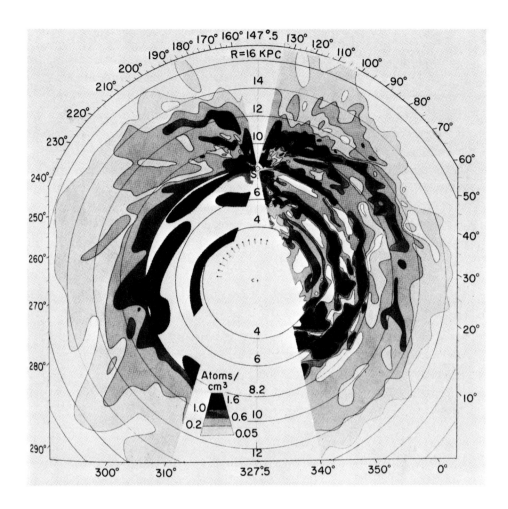

Spiral pattern of our galaxy. Hydrogen gas appears to follow a spiraling pattern within our galaxy, as indicated in the plan view of the clouds that have been detected by radio telescopes. Many, but not all, features of this pattern coincide with features of the spiral pattern discerned from the emission nebulosities and the clusters of young stars. The sun is marked "S" and the center of our galaxy is at the point "C." (One KPC is approximately 3.2 light years.) This diagram was prepared at the Leiden Observatory in Holland. (*Reproduced from O. Struve and V. Zebergs: Astronomy of the Twentieth Century, New York, 1962*)

relative amount of each type of radiation is determined by the speed of the electrons and the strength of the magnetic field.

Neutral gas cannot produce synchrotron radiation, so the nucleus must contain ionized clouds. A small cloud of hydrogen atoms has been discovered just outside the nucleus, violently rotating and expanding from the center. High above the plane of the Milky Way lie a number of similar clouds, and some of them appear to be streaming *toward* the center. Perhaps there is a circulation inward from above and jetting outward along the plane of the Milky Way.

Whatever their cause and their significance, such clouds are fairly common in the universe. Many galaxies show similar activity, some with far greater violence, and the most violent of these galactic nuclei are remarkably similar to quasars, the most intense sources of radiation discovered thus far.

If Ambartsumian was correct in guessing that matter is streaming from the centers of galaxies, we may be witnessing the births of vast stellar systems. If that is true, astronomers stand at the brink of another revolution.

EPILOGUE

In 1942, at the time of Baade's attempts to detect individual stars in the central portions of the Andromeda Nebula, another astronomer at the Mount Wilson Observatory put the finishing touches on a paper describing extraordinary activity in the nuclei of some spiral nebulae. He found a small fraction of intermediate spirals showing a brighter nucleus than the average; and the nucleus contained a cloud—or clouds—of rapidly expanding gas. The spectrum of the gas showed that it was of low density and high temperature, similar to the gas in a planetary nebula, but most startling was the fact that gas was leaving the nucleus at several thousand miles per second, and was destined to escape into intergalactic space long before these galaxies could complete another turn.

The author was Carl Seyfert. At first his paper attracted the half-hearted attention often accorded to curiosities. Clearly something unexpected was occurring in galaxies, but no one could put these events into a broader context.

These galaxies became known as "Seyfert galaxies," and they have been recognized as a possible link between the activity at the center of our own galaxy and the catastrophic events seen in other galaxies: one neighbor of ours, the giant elliptical galaxy M87, has ejected a jet from its nucleus (plate, page 281); the galaxy M82 seems to have turned itself inside out (plate, page 282).

Closer to the sun and within the Milky Way is the Crab Nebula, evidently the debris of a supernova explosion in 1054 A.D. The filamentary structure and the peculiar light of this object (plate, page 171) are similar to those of M82. In both cases, much of the light is generated by electrons spiraling their way down streamers of a magnetic field. But the similarity stops there, because the Crab Nebula is tiny compared to M82 and it contains relatively little mass. In the debris, visible but undetected until a few years ago, blinks a "pulsar." This evidently is the ash of the original star. It is a tiny star of neutrons—fundamental particles so crowded that they cannot form full-blown atomic nuclei. The star rotates one hundred times per second, twisting a magnetic field about it like a pirouetting dancer with a scarf and sending out pulses of radio radiation. The neutron star contains only a small fraction of the original star's mass; the remainder consists of the nebula and is on its way to the edges of the Milky Way.

These recent glimpses of galactic violence have raised questions which go to the basis of current speculations on the origin of galaxies and the expansion of the universe.

Until recently, most astronomers assumed that galaxies were formed by the condensation of a chaotic medium that had pervaded the universe. Primordial chaos was the starting point of ancient Greek and

---

The nuclear jet in M87. This giant elliptical galaxy (see also plate, page 61) is evidently the seat of violence. The light of the jet is blue and it is highly polarized, as though it were produced by fast electrons streaming along a magnetic field. This galaxy is a strong source of radio radiation also. (*Lick Observatory, University of California*)

An exploding galaxy, M82. This galaxy appears to have ejected enormous quantities of matter from the central region. The vertical filaments are produced by hydrogen, so their light is red and it may be distinguished from the starlight with colored filters. (*Hale Observatories*)

Hebrew traditions, and it has been incorporated into most of the modern cosmologies. No one pretends to understand the details of the condensation process, but most astronomers agree that no matter how it actually occurred, such a condensation would naturally lead to galaxies in rapid rotation, in analogy with the formation of the solar system from an extended swirl of interstellar gas and dust. (*Cf.* Chapter 13.)

When Baade announced the existence of two stellar populations within the Andromeda Nebula and they were discovered within our own galaxy, astronomers quickly developed an explanation for the fact that the older population rotated much more slowly than the younger population. The old population was said to have been born early, when the dusty gas of the Milky Way retained its primitive spherical form and before it had collapsed into a rapidly rotating disk. This explanation seemed to be nicely confirmed by the chemical differences between the new and the old generations: the older stars contained fewer of the heavy complex atoms, presumably because these atoms were formed in large quantities after the birth of the older population.

But there is an aspect of the evolution of galaxies that is not so easily explained. If the rapidly rotating disks of galaxies are formed *after* the slowly rotating spherical collections in the center, why are the disks so much *larger* than the central portions?

An intuitive picture of the condensation of a galaxy might proceed the following way. The gas and dust would start in a roughly spherical form and contract, rotating more rapidly and flattening. Small condensations in the cloud would produce stars and clusters of stars; those which were formed early would consist of a spherical, slowly rotating system about the center of the galaxy—Baade's population of type II.

After the formation of the first generation of stars in a spherical system, the residual gas would continue contracting into a disk. It is not easy to see how continued contraction, in itself, could lead to a disk that is larger than the spherical cloud from which it formed. Some astronomers have postulated a series of drastic events preceding the second generation. They suppose that the original stars acted like great stewing pots and cooked the simple elements (hydrogen and helium, for example) into heavier and more complex atoms (oxygen and iron, for example). And they suppose that these stars then sprayed their material back into space, making it available for the second generation. In this way they hope to account for the difference of chemical composition.

But what of the difference between the sphere of the first generation and the disk of the second? And why does the second generation extend well beyond the spatial limits of the first? According to one explanation, the "cooked" material was ejected violently and it spread into a much larger spherical volume. Then it contracted into a disk, and subsequently it produced stars.

There is a remarkable feature of galaxies which may be evidence that galaxies have not condensed from a chaotic medium pervading the

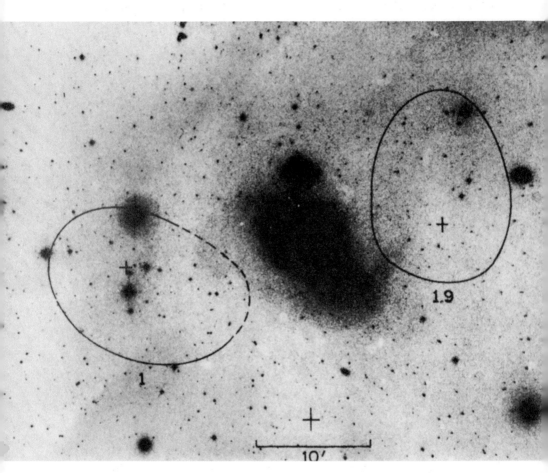

A source of radio signals, N.G.C. 1316. This peculiar galaxy shows two sources of radio signals, one on each side, as though a cloud had been ejected in both directions along a magnetic field. This is not an unusual configuration among radio sources. (*Reproduced from the* Astrophysical Journal, *Vol. 147* [*1964*], *p. 44*)

universe: there appears to be a very well defined maximum to the size of galaxies.

A continuous scale of sizes seems to connect the small globular clusters to the dwarf galaxies, to the companions of the Andromeda Nebula, to our own galaxy, and finally to the giant ellipticals. However, all of the large elliptical galaxies appear to be of about the same size, and this consistency contradicts some people's intuitive ideas about the random condensation of galaxies. (It is true that *stars* also show a well-defined maximum, and they are supposed to have formed by random condensation, but in their case we assume that the largest stars have already died. All we see now are the smaller ones, which have managed to survive. Rather than suppose that this could have happened to the galaxies, whose behavior is not very similar to that of an individual star, some astronomers seek drastically different origins for the two types of objects.)

To be inside the nucleus of a Seyfert galaxy would be a remarkable experience; the sky would be ablaze. In March 1970, Martin Schwarzschild of Princeton University supervised the flight of a thirty-six-inch telescope in an unmanned balloon to a height of eighty thousand feet. The purpose was to look for fine detail on the face of the planets and within galaxies. A Seyfert galaxy (N.G.C. 4151) was observed whose nucleus appeared too small in the telescope to be distinguished from a star. From this fact, Schwarzschild concluded that its diameter is less than 17 light years, or about four times the distance of the star nearest the sun. Yet the brightness of this nucleus is that of two billion suns!

It has been suggested that this nucleus is a cluster of ten billion stars moving among each other at one thousand miles per second. In such a cluster, a collision would occur every few months, and such collisions might account for a large fraction of the light from the nucleus. The light would vary, flickering up after a collision and then sinking down as the explosion fades. Variations have in fact been observed, but they appear to occur every year or so, and this may not be frequent enough to be consistent with the suggestion.

From the average interval of one year between the changes of brightness in this nucleus, we conclude that if the nucleus is a single object it cannot be more than one light year across. If it were much

The peculiar barred spiral N.G.C. 2685. There is no doubt that this is a galaxy, but its configuration is a total mystery. Evidently the central ellipsoid is surrounded by helical streamers of absorbing matter. (*Hale Observatories*)

larger, its parts could not vary in synchronism with one another as seen from the earth. Signals could not travel among all of its parts in less than a year, and if the parts were not connected by signals they could not vary in a coherent fashion; most of the elements would cancel each other out; we would only notice a tiny residual flickering rather than a substantial change of brightness.

This argument is now used quite generally in astronomy: An object which shows large variations with a period of T days cannot be much more than T light days in diameter. (The corresponding distance is cT, where c is the speed of light, 186,000 miles per second.)

As an analogy, consider a large chorus singing in the dark, where sound provides the only means of communication among the singers. If the chorus were spread over a distance of ⅕ mile, say, the sound would require one second to travel from one edge to the farthest point. The conductor would have a hard time keeping the chorus in synchronism to better than ½ second if he stood in the center.

To make the analogy closer, suppose the conductor wishes the chorus to sing a short burst of music, say a single note held for ¹/₁₀ second. If he stands in the center of the chorus, the shortest note heard by a listener standing outside will be a full second—the difference of travel time from the far and near points.

In the same way, astronomers argue that if a light source can vary with a period of a year, it cannot be more than one light year in diameter. This argument leads to the conclusion that the nucleus of the Seyfert galaxy examined by Schwarzschild is ten times smaller than the light set by the balloon observation.

This argument has a weakness. Note that if the conductor wishes to get a shorter note out of his chorus he can place himself on the side of the chorus opposite to the listener. Then, if each singer follows the conductor's voice signals, and if the conductor is not too close to the chorus, the listener will hear them all at precisely the same time; they will sound in synchronism because the singers closest to the conductor and farthest from the listener will sing first, etc. So, by a proper arrangement of the geometry, it would be possible to get a very short note out of the chorus even though the singers were spread out.

Perhaps the most remarkable property of the Seyfert galaxies is this. If we imagine the outer, spiral portion to be removed, the nucleus

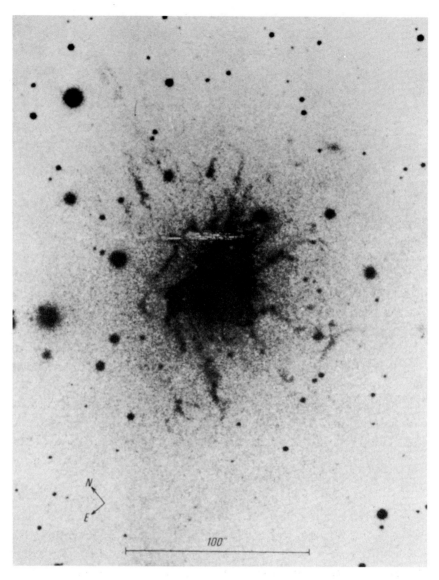

An explosion? This object, N.G.C. 1275, is evidently an external galaxy which has suffered an explosion; perhaps it is similar to M82 (plate, page 282). Its filamentary structure is remarkably like that of the Crab Nebula, although on a much larger scale. (*Reproduced from the* Astrophysical Journal, *Vol. 159* [*1970*], *p. 1152*)

would appear to be indistinguishable from a very distant "quasar." (See plate, page 290.)

The quasars are a newly recognized component of the universe. Discovered in 1963, they are now thought by most astronomers to be the most luminous objects in the sky—and therefore detectable to the greatest distances.

A quasar is defined by two properties: its image is starlike, or it contains a dominant, starlike component; its spectrum displays a redshift much larger than those of ordinary stars in our galaxy. Some quasars emit strong radio signals; these are the QSS's (quasi-stellar sources). Some quasars, although displaying all the other features of QSS's, do not emit strong radio signals; these are the QSO's (quasi-stellar objects). All quasars show a distinctive color—very strong ultraviolet light—and this property facilitates their discovery.

If the redshifts result from recessional velocities near the speed of light, two alternative explanations are open. Either the quasars are at great distances and their motions reflect the general expansion of the universe, or they are relatively close and their motions are the consequences of violent impulses. (They might, for example, have been ejected from the nuclei of nearby galaxies.) Thus far, a definitive distinction between these alternatives has not been possible, but the majority of astronomers believe that they are at great distances.

On this basis, a few things can be said with fair certainty about the quasars. They emit light that resembles the light of galactic nuclei, but they radiate a thousand times as much as the brightest galaxies known. Some are accompanied by faint wisps, suggesting that they have ejected clouds of gas. From the distances and the speeds of such clouds, a minimum age may be estimated for the quasars: a million years.

Quasars display light variations like those of the Seyfert galaxies, so they cannot be much larger than a few light years. That is the enigma of the quasars: How can they produce a thousand times the light of an entire galaxy for a million years and yet be smaller than the distance to the sun's nearest neighbor?

When, in the nineteenth century, Norman Lockyer suggested that stars derive their energy from the infall of matter, he did so for a very good reason: enormous quantities of energy can be obtained this way. Gravitational energy, the energy obtained by allowing an object to fall

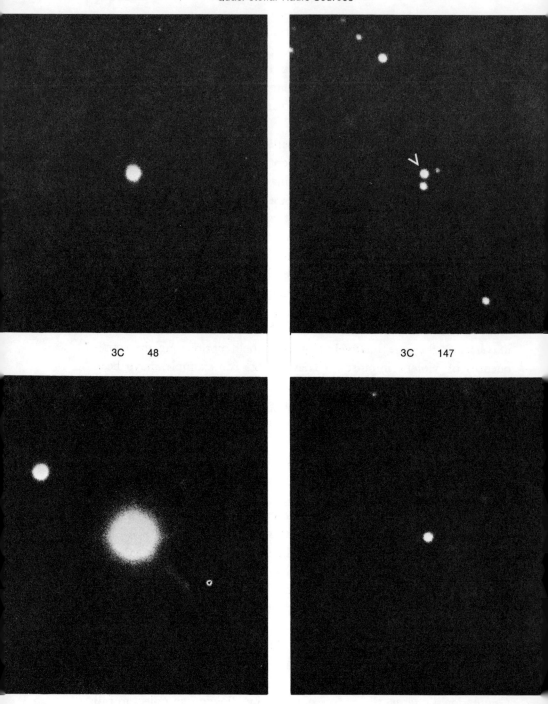

3C 48

3C 147

3C 273

3C 196

in a gravitational field, is much more easily accumulated than electrical or chemical energy because gravitational fields penetrate matter and can exercise their influence to the edge of the visible universe. Electrical forces and chemical forces, which are just the electrical forces among atoms, are shielded and neutralized by the mixture of positive and negative charges, so they cannot be felt at great distances. But gravity, unlike electricity, does not possess positive and negative charges; it cannot be neutralized.

To see how great is the energy available from gravity, consider the following. If an object of mass, m, falls to the surface of a star and attains a velocity, v, the energy of infall is $\frac{1}{2}mv^2$. According to Einstein's relation, the energy available from annihilation of the matter and its total conversion into light would be mc^2, so the gravitational energy can approach the annihilation-energy if the particle falls inward with a speed near the speed of light. This speed can be achieved if the star is massive and its radius is small.

Suppose matter continues to stream into a very dense star, accelerating and heating as it falls toward the surface. As the mass of the star increases, additional matter piles up on the surface and the inner portions are compressed. As the mass of the star continues to pile up, each added bit presses harder because the gravitational pull at the surface of the star is greater. Finally, the interior of the star may be unable to resist; it would shrink as each particle lands on its surface. But when the star shrinks, the gravitational pull at its surface is increased even more, so the material that comes in later is just that much heavier. The process runs away with itself—as the matter continues to stream inward, the object collapses. This may be the stage at which it looks like a quasar.

Four quasars. These objects are many times brighter than our own galaxy, but they lie near the limits of the observable universe, hence their apparent faintness. Quasars are identified by the redshift of their spectra, produced by rapid motion away from our galaxy. Note the wisp near the object 3C273. Evidently it was ejected from the nucleus of the quasar, perhaps in a manner similar to the wisp ejected from the galaxy M87 (plate, page 281). (*Hale Observatories*)

Now a very strange thing can happen. If there is sufficient matter within the collapsing star, there will be an observable "time-dilatation" of the interior with respect to the exterior. Suppose a clock is dropped toward this object. To an observer falling with it, the clock will appear to behave quite normally, but to an observer who remains outside, the rate of the clock will decrease as it falls deeper into the gravitational pit. (This apparent change in the rate of the clock was predicted by Einstein's General Theory of Relativity and it has been confirmed in the laboratory and by observations of the sun, so I will simply declare it to be a fact of nature, although I cannot claim to "understand" it in terms of personal experience.)

Not only will the rate of the clock appear to decrease as we watch it fall, but all the atoms in its neighborhood will radiate more slowly; their light will appear to turn red and become feeble, as though the voltage had been turned down in an electric light bulb. Ultimately, the clock will appear nearly to stop, and it will almost vanish because very little light can escape from it. Peering down after the clock, we will see nothing but blackness—a "black hole" will have engulfed the clock.

The matter which falls into a black hole will fade from view and will disintegrate in the stress of gravity. Given a sufficient time, its radiation will become too feeble to be detected against the background of ordinary star light in the sky. Although the matter of the black hole cannot re-escape as matter, it will be converted into a relativistic fluid of radiation-matter and some of it may manage to escape in the form of feeble red radiation. This, however, is a rudimentary view of black holes. Some cosmologists suggest that the future theory will indicate that black holes are the links to another realization of matter; they may be the passageways to another universe, just as the manholes of Paris lead to a world beneath the street.

Here is another peculiarity of the black hole: both the insider and the outsider think that the other has turned red and slowed nearly to a halt; neither senses a change in himself. If we rode in with the clock, accelerating toward the speed of light as we descended, we would see the stars behind us turn red and fade from our vision. Time on the outside would appear to slow down for us.

How would it end? Unless a new process occurred—a process that has not yet been discovered or whose operation in the black hole has not yet been recognized—the collapse would continue indefinitely. Even if the stream of infalling matter were cut off from the outside, the matter that had already fallen in would be trapped and cut off from the universe that we know.

Although I am convinced that black holes exist and I have an intuitive sense of their properties, I have not yet learned to describe them mathematically and to make their properties seem coherent to other people. That will be the task of John A. Wheeler, of Princeton, and other cosmologists who have helped develop the concept. My faith in the existence of these holes is based on my study of astronomy and, in particular, of relativity. When I discuss such subjects with my friends and family, I feel as though I were trying to convince them of the existence of God, and I have come to expect their quizzical expressions. I suppose my faith in black holes is one that has grown from my lifetime in science.

Black holes may be detected through their gravitational influence on objects in their neighborhood. Astronomers hope to track them down within dense clusters of stars or the interiors of elliptical galaxies, but the hunt is going to be a difficult one because their observable effects may be subtle.

Quasars, and the thoughts they generate, have led Fred Hoyle to advocate what he has called a "radical departure" from the steady-state theory in its original form.

Hoyle now proposes that matter is created, not uniformly throughout the universe, but in regions of high density and intense activity, such as a developing black hole. Perhaps, he says, a new process starts up when the matter becomes sufficiently compressed and a counter-pressure stops the collapse and sends the matter outward again.

The outward-streaming matter from a reopened black hole might produce effects such as are now seen in quasars and in the nuclei of galaxies. It would flow as if from nowhere.

The need to postulate a new process to reverse the collapsing matter and disperse it again in the expanding universe is not peculiar to Hoyle's theory. Such a process would also be required if we suppose that the big bang is a cyclic process and that the universe as a whole has bounced outward from a collapse.

At present, it appears possible that the center of our own galaxy has been the collection point for matter that has fallen in from the Milky

Dust clouds emerging from the nucleus of M31 may indicate the ejection of matter. (*Hale Observatories*)

Way and is now in the process of bouncing back into the outer reaches of space. Perhaps, on a larger scale, we are seeing the same thing in the nuclear jet of M87 and in the disruption of M82. (See plates, pages 281 and 282.)

Many clusters of galaxies are dominated by a large elliptical galaxy emitting strong radio radiation as though its core were the seat of violence. Some astronomers have inferred that matter, streaming out of such galaxies, has produced the other members of the cluster. (Strings of galaxies, arranged in astonishing and very "artificial" patterns, suggest that galaxies can be produced in groups.) If we assume that the quasar activity is an enhanced version of the activity in Seyfert galaxies and the nuclei of giant ellipticals, we can imagine that they might give birth to enormous clusters of galaxies.

In effect, Hoyle suggests that we combine the concept of a steadily expanding universe, in which matter is created continually, with the concept of big bangs, in which the created matter and radiation emerge explosively from regions of extremely high density. According to this new picture, contractions and re-expansions take place at scattered points within the present universe.

This new conception would provide a theoretical connection among many of the diverse phenomena of the universe. It has another attractive feature: we might find a source for the three-degree background radiation of the universe in these explosive centers. (See Chapter 19.) Just as it had been assumed that the radiation might be the residue of the original heat radiation of the big bang, so it would now seem possible to explain the radiation as the product of black holes.

The old view of galaxies as quiescent swirls of stars, gas, and dust is giving way; signs of violence are everywhere.

When quasars and black holes find their places in cosmology, and their roles are understood, astronomers will have written the most complete revision of the creation story since the days of the Ancients.

GLOSSARY

Astrologer—A person who attempts to understand the lives of people through a study of the stars.

Astronomer—A person who attempts to understand the stars and planets through measurement and theory. Call him an astrologer and he will smile.

Binary star or *double star*—A pair of stars in motion about a common center. Many such stars appear as double in a telescope, but often the pair is so close together that its duplicity can only be detected indirectly; e.g., with a spectograph.

Cepheid variable—A star whose brightness varies regularly with a fairly well-defined pattern. The variations are due to pulsations in the size and temperature of the star, and the period is closely related to the luminosity of the star; hence these stars are used for distance measurement.

Cluster—A group of stars that were formed together and are held together by gravity. *Galactic* clusters are found along the Milky Way, and they are rather loose groupings of hundreds of stars. *Globular* clusters are spread throughout a roughly spherical volume with a concentration at the center of our galaxy; they contain millions of stars.

Cosmology—Study of the structure and evolution of the universe.

Cosmogony—Study of the origin of the universe, and particularly of the solar system.

Doppler Effect—The observed change in the pitch of a vibrating signal (such as a sound wave or a light wave) when emitted by a source that is in motion with respect to the observer. In starlight, it is detected with a spectrograph.

Galaxy—A gravitationally bound unit of stellar aggregation; an "island universe"; a complete stellar system. "The Galaxy" is reserved for the system in which we live and which we see as the Milky Way. Loose clusters of galaxies have also been discovered.

Hertzsprung-Russell (or *H-R*) *diagram*—A plot on which each star is located by its luminosity and by its color or spectral type.

Luminosity—The total radiation emitted by a star, hence a property of the star and not of the observer's location.

Magnitude—Numerical measure of the brightness of an astronomical object. The twenty brightest stars were assigned to the *first* magnitude by the Ancients, and the remaining visible stars were assigned to five further magnitudes. Modern techniques have permitted the definition of many different types of magnitudes, depending on the color of the light that is selected for measurement.

The *absolute* magnitude of a star is the magnitude which it would have at a standard distance. Hence it is a property intrinsic to the star, as is luminosity.

Milky Way—The ancient name of the band of diffuse starlight circling the sky. Today, with the recognition that the appearance of the Milky Way is produced by our location within the galaxy, these names are often interchanged rather casually.

Nebula—A diffuse-appearing object. *Bright* nebulae shine by their own light; *dark* nebulae are dark regions etched on a stellar background. Dark nebulae are formed of dust. Bright nebulae within our galaxy are formed of luminous gas. The term *extragalactic nebula* is often applied to galaxies other than our own.

Nova—A star whose brightness increases by a factor of a hundred or more (often very much more) within several days and then decreases to normal within a year or so. Ordinary novae appear by the dozens each year in a galaxy, while *supernovae,* which become nearly as bright as an entire galaxy, occur only once a century or so. The residue of a supernova is an expanding nebula and a pulsar—this, at least, is the evidence of the Crab Nebula and its pulsar.

Parallax—The apparent elliptical motion of a nearby star against the background of more distant stars during the annual motion of the earth about the sun. Measures of parallax permit direct determination of the distance, but the method is limited in practice to the closer stars.

Planet—Originally, this word referred to the "wandering stars" which were observed to move along the sun's path through the sky. Now it applies generally to any cold body which is in orbit about a star. Jupiter is nearly the largest planet which can exist, because additional matter would cause the interior to contract.

Populations of type I and type II—A designation invented by Walter Baade to distinguish the stars confined to the disk of spiral galaxies from those whose orbits take them into a more spherical distribution. Observationally, the types are also distinguished by the color of the brightest stars and by their chemical composition. Theoretical work suggests that the population I stars are much younger than those of population II.

Proper motion—The apparent motion of a star across the sky; it reflects the difference of the motions of the sun and the star through space.

As most stars within a galaxy move with comparable speeds, the most distant stars show the smallest proper motions.

Radial velocity—The speed with which a star or galaxy recedes from the sun. It may be measured through the Doppler Effect in the light of the object.

Redshift—A reddening of the light of galaxies. It increases with distance, and when interpreted as a Doppler Effect it implies a velocity of recession that is nearly proportional to the distance.

Spectral type—A classification assigned to a star on the basis of the appearance of its spectrum. Numerous systems have been devised empirically since the closing decades of the nineteenth century.

Star—Originally, this word applied to any bright object in the sky, with the exclusion of the sun and the moon, although it is now reserved for self-luminous bodies which are held together by gravity and held apart by interior pressure.

BIBLIOGRAPHICAL NOTES

☼

PART I

My discussion of ancient astronomy is incomplete and one-sided because I chose to focus on Greek Epicureanism. A more balanced treatment may be found in *The Fabric of the Heavens,* by Stephen Toulmin and June Goodfield (New York, 1961). Epicurean philosophy and science are set forth in Lucretius' poem "The Way Things Are," and its recent translation by Rolfe Humphries (Bloomington, 1968) is my choice as the most delightful piece of ancient philosophical literature available in English.

Alexandre Koyré has written a number of important books on the history of science; his *From the Closed World to the Infinite Universe* (New York, 1957) describes man's attempt to cope with the concept of the universe between the fifteenth and the eighteenth centuries.

Arthur Koestler's book *The Sleepwalkers* (New York, 1959) recounts the lives of Copernicus, Brahe, and Kepler. It is a lively book and is tinted by the color of Koestler's style. For me, this is part of its charm. Kepler's response to the telescopic discoveries of Galileo are recorded in *Kepler's Conversation with Galileo's Starry Messenger* (New York, 1965), by Edward Rosen.

Nothing that has been written about Galileo can surpass what he, himself, has written. Excerpts from his astronomical writing are available in Stillman Drake's *Discoveries and Opinions of Galileo* (New York, 1957). A brief biography, which places his work in context, is Laura Fermi and Gilberto Bernardini's *Galileo and the Scientific Revolution* (New York, 1961).

One of the first astronomical biographies I read as a boy was Zsolt de Harsanyi's *The Star-Gazer* (New York, 1934). Galileo came alive to me through this book.

The significance of Copernicus, and the broader question of the nature of revolution in science, are explored by Thomas Kuhn in *The Copernican Revolution: Planetary Astronomy in the Development of Western Thought* (Cambridge, Mass., 1957).

Isaac Newton's most important book, *The Mathematical Principles of Natural Philosophy* (also known as *The Principia*), is nearly inscrutable to the layman, but an introduction to the underlying ideas may be found in I. Bernard Cohen's *The Birth of a New Physics* (New York, 1960). Newton's life is described by E. N. da C. Andrade in *Sir Isaac Newton* (New York, 1958), and the best psychological interpretation of Newton is Frank E. Manuel's *Portrait of Newton* (Cambridge, Mass., 1968). Some of Newton's letters to Richard Bentley may be found in *Theories of the Universe* (New York, 1957), edited by Milton Munitz. This book contains excerpts from a wide range of historical sources.

Thomas Wright's book *An Original Theory, or New Hypothesis of the Universe* has recently been reprinted (New York, 1971), and his *Second, or Singular Thoughts upon the Theory of the Universe* has been edited from the manuscript by Michael Hoskin (London, 1968).

Kant's theories of the solar system and the universe, and the fascinating relationships among his theories and those of Wright, Lambert, and Laplace, are described by William Hastie in *Kant's Cosmogony* (Glasgow, 1900). His theory is also available in a revised edition, with an introduction and appendix by Willy Ley (New York, 1968).

As far as I know, the variorum translation of Laplace's cosmogony is given here for the first time. It is based on a scrutiny of the French texts of his *Système du Monde* by Miss Sylvia Boyd, and I am indebted to Mrs. Françoise Mertz for assistance in the translation.

PART II

For my interpretation of William Herschel's work, I have drawn heavily on Michael Hoskin's book *William Herschel and the Construction of the Heavens* (London, 1963). Herschel's personal life may be glimpsed through two collections of letters and journals: *The Herschel Chronicle*, edited by Lady Constance Ann Lubbock (New York, 1933); *Memoirs and Correspondences of Caroline Herschel*, edited by Mrs. John Herschel (London, 1876). John Herschel's poem quoted in Chapter 14 was published in *Essays from Edinburgh and Quarterly Reviews*, Sir John Frederick William Herschel (London, 1857). Sir John's life is the subject of Günther Buttmann's *Shadow of the Telescope* (New York, 1970). Lord Rosse's life has not yet been treated in detail.

PART III

Agnes Clerke's *A Popular History of Astronomy During the 19th Century* (London, 4th edn., 1902) is a comprehensive narration that is generally considered to be *the* book on the nineteenth century.

As far as I know, the only book treating the twentieth century with anything like the same completeness is Otto Struve and Velta Zebergs' *Astronomy of the 20th Century* (New York, 1962). I have the impression that this book was aimed at the intermediate-level college student, but I know that it provides valuable perspective to the research astronomer, and I suspect that the uninitiated would find portions of it stimulating also.

Alfred Wallace's discussion of the structure of the universe and the likelihood of extraterrestrial life is found in his book *Man's Place in the Universe* (New York, 1904). This book gives an excellent synthesis of astronomical knowledge at the close of the nineteenth century—as seen through the eyes of a biologist with a clear conviction of God's purpose.

The British astronomers Arthur S. Eddington and James H. Jeans each wrote several popularizations of astronomy during the 1920's and 1930's. Their books are largely outdated, but they endure as models of fine exposition. Jeans's philosophical discussions were an attempt to incorporate modern physics into philosophy. Eddington's book *Fundamental Theory* (Cambridge, Eng., 1946) is regarded by some as a sign of Eddington's having ascended to a mysticism of numbers during his later years.

Among the popular treatments of astronomy from the modern perspective, my favorite is George Abell's *Exploration of the Universe* (New York, 1964).

An introductory book which focuses on the nature of our galaxy is Bart J. and Priscilla F. Bok's *The Milky Way* (Cambridge, Mass., 1957). This book dives into subjects that I have merely skimmed.

My candidates for the three most important books on galaxies are: Edwin Hubble: *The Realm of the Nebulae* (New Haven, 1936); Harlow Shapley: *Galaxies* (Cambridge, Mass., rev. edn., 1961); Walter Baade: *Evolution of Stars and Galaxies* (Cambridge, Mass., 1963). The first of these is already a classic of science, the second is part of the Harvard Series on Astronomy (aimed at beginning students), and the third is a discursive book that is cherished by professional astronomers as well as students because Walter Baade rarely lectured or published.

Photographs of galaxies may be purchased from The Hale Observatories (address correspondence to the California Institute of Technology Bookstore, 1201 East California Boulevard, Pasadena 91109); The Lick Observatory, University of California, Santa Cruz 95060; The Yerkes Observatory, Williams Bay, Wisconsin 53191.

For the story of George Ellery Hale and the development of the Mount Wilson Observatory (recently renamed the Hale Observatories along with the Mount Palomar Observatory), I drew on the biography by Miss Helen Wright, *Explorer of the Universe* (New York, 1966).

Harlow Shapley's reminiscences are quoted from *Through Rugged Ways to the Stars* (New York, 1969), and this is but one of many in the spirited herd of Shapley's books.

The popular books of Fred Hoyle, a British astrophysicist who has also written science fiction and mystery stories, are read faithfully by many astronomers, because they provide a cohesive picture of speculations such as are not ordinarily found in professional journals. Hoyle's book *Galaxies, Nuclei and Quasars* (New York, 1965) and *The Violent Universe* (New York, 1969) by Nigel Calder both discuss quasars and exploding galaxies. For pulsars, see the article by A. Hewish in *Annual Reviews of Astronomy and Astrophyics,* Vol. 8 (Palo Alto, 1970).

The concept of "black holes" and its application to astronomy are still in the process of elucidation. The fullest discussion is in the January 1971 issue of *Physics Today.* It was prepared by Remo Ruffini and John A. Wheeler, both of Princeton University. At the center of the concept is gravity, as expressed by the General Theory of Relativity. An excellent introduction is Peter G. Bergmann's book *The Riddle of Gravitation* (New York, 1968) and, to my relativistically naïve mind, the best general discussion of relativity is Albert Einstein and Leopold Infeld: *The Evolution of Physics* (New York, 1938).

John Ziman has provided a brilliant definition of science and its distinction from, say, religion in *Public Knowledge* (Cambridge, Eng., 1968). Stephen Toulmin's *Foresight and Understanding* (New York, 1961) is a small book which I invariably recommend to those interested in the attributes of a good scientific theory. A large book, and one which has had a substantial influence on my own philosophy of science, is *Personal Knowledge* (Chicago, 1958) by Michael Polanyi. The apparent antithesis implied by the titles chosen by Ziman and Polanyi is apparent only, and it reflects the recent recognition that science has, of necessity, both personal and public aspects. Scientists are releasing themselves from the straightjacket of purely rational analysis; some have come to view themselves as poets attempting to test their poems—or something along those lines.

INDEX

Page numbers in italic type indicate illustrations.

A NOTE ABOUT THE AUTHOR

Charles Allen Whitney was born in Milwaukee, Wisconsin, in 1929. After re-
ceiving an S.B. in physics from the Massachusetts Institute of Technology, Whitney completed his A.M. and Ph.D. in astronomy at Harvard University. He has
been a physicist at the Smithsonian Astrophysical Observatory since 1956, and is
also a professor of astronomy at Harvard. No less a writer than a scientist, Whitney is recipient of a Guggenheim Fellowship for his projected biography of Edwin Hubble. He is married, the father of five children, and lives in Weston, Massachusetts.

The text of this book was set on the Linotype in Garamond, a modern rendering of the type first cut by Claude Garamond (1510–1561). Garamond was a pupil of Geoffroy Troy and is believed to have based his letters on the Venetian models, although he introduced a number of important differences, and it is to him we owe the letter which we know as old-style. He gave to his letters a certain elegance and a feeling of movement that won for their creator an immediate reputation and the patronage of Francis I of France.

Composed and bound by H. Wolff Book Manufacturing Co., Inc., N.Y.
Printed by Halliday Litho, West Hanover, Mass.
Typography and binding design by The Etheredges.